Bas van Fraassen's Approach to Representation and Models in Science

SYNTHESE LIBRARY

STUDIES IN EPISTEMOLOGY, LOGIC, METHODOLOGY, AND PHILOSOPHY OF SCIENCE

VOLUME 368

For further volumes:
http://www.springer.com/series/6607

Wenceslao J. Gonzalez

Editor

Bas van Fraassen's Approach to Representation and Models in Science

 Springer

Editor
Wenceslao J. Gonzalez
Faculty of Humanities
Department of Humanities
University of A Coruña
Ferrol (A Coruña), Spain

ISBN 978-94-007-7837-5   ISBN 978-94-007-7838-2 (eBook)
DOI 10.1007/978-94-007-7838-2
Springer Dordrecht Heidelberg New York London

Library of Congress Control Number: 2013956816

Printed on acid-free paper

Springer is part of Springer Science+Business Media (www.springer.com)

Contents

Prologue

New Analyses on Representation and Models

Wenceslao J. Gonzalez

Bas van Fraassen is a key figure in contemporary philosophy of science, as the prestigious Hempel Award explicitly recognizes.[1] He has developed a new approach to representation and models in science. His views on scientific representation offer new ideas on how it should be characterized, and his conception of models shows a novelty that goes beyond other empiricist approaches of recent times. Both aspects — the characterization of scientific representation and the conception of models in science — belong to a deliberate attempt to forge a "structural empiricism," an alternative to structural realism based on an elaborated version of empiricism.

This book follows several steps in dealing with van Fraassen's approach to scientific representation and models in science. First, the volume offers the philosophical coordinates of his views on science, in general, and on scientific representation and models, in particular. Second, there is a renewed attention to the structural empiricism on models and representations, which includes a new contribution made by van Fraassen and a reflection on his approach. Third, the attention shifts to the relation between models and reality, where the complexity of his conception is considered in detail. Fourth, there is an examination of scientific explanation and epistemic values judgments, which includes another contribution by the author studied here.

Each one of these steps involves several papers. (1) *Philosophical Coordinates* are considered in three chapters: "On Representation and Models in Bas van Fraassen's Approach," Wenceslao J. Gonzalez (University of A Coruña); "Scientific Activity as an Interpretative Practice. Empiricism, Constructivism and Pragmatism," Inmaculada Perdomo (University of La Laguna); and "Models and Phenomena: Bas van Fraassen's Empiricist Structuralism," Valeriano Iranzo (University of Valencia). (2) *Models and Representations* are in the direct focus of two chapters: "The Criterion of Empirical Grounding in the Sciences," Bas van Fraassen (San Francisco State

---

[1] The Philosophy of Science Association has given this recognition to him. Bas van Fraassen received this inaugural award at the PSA meeting in San Diego, California, on 17 November 2012. In this meeting he expressed his satisfaction for receiving an Award with the name of Carl Gustav Hempel.

University); and "On Representing Evidence," Maria Carla Galavotti (University of Bologna).

Thereafter, the book has another two groups of chapters, each with two papers. (3) *Models and Reality* includes the topics discussed in "Scientific Models and Abduction: The Role of Non Classical Logic," Ángel Nepomuceno (University of Seville); and "The View *from Within* and the View *from Above*: Looking at van Fraassen's Perrin," Stathis Psillos (University of Athens). (4) *Scientific Explanation and Epistemic Values Judgments* are the leitmotiv of two contributions: "Explanation as a Pragmatic Virtue: Bas van Fraassen's Model," Margarita Santana (University of La Laguna); and "Values, Choices, and Epistemic Stances," Bas van Fraassen (San Francisco State University).

According to this configuration of the contents of the book, Part I offers a general framework in the first paper, "On Representation and Models in Bas van Fraassen's Approach." It embraces key elements of the philosopher analyzed and his theses on the central topics of this volume. Wenceslao J. Gonzalez presents the salient traits of the intellectual trajectory of Bas van Fraassen, which is followed by the scrutiny of the philosophical context of representation and models in science. Thus, the realms of analysis of representation and the features of representation in science are explicitly considered. They involve a combination of empiricism and pragmatism in van Fraassen's approach, which moves from "constructive empiricism" to "structural empiricism." These conceptions are alternatives to some accounts of scientific realism, mainly those versions of realism in science that were more influential when *Scientific Image* was published (van Fraassen 1980) and when *Scientific Representation* came out (van Fraassen 2008).

Through the study of van Fraassen's approach to scientific representation and models, in Chap. 1 Gonzalez emphasizes several aspects of the proposed conception on representation as activity and the elaborated view of the role of models. These aspects illustrate that van Fraassen's approach is mainly logico-epistemologic, where the focus is on basic science rather than applied science. Ultimately, his proposal on representation and models in science is a combination of a set of philosophical elements: (i) a pragmatic position regarding language; (ii) an empiricist epistemology; (iii) constructivism on methodology of science; (iv) a pragmatist ontology; and (v) a special emphasis on cognitive values within the axiological realm.

Following van Fraassen's structural empiricism, representation should be where the realist likes to put truth, and representation is triadic (a representational structure, a target, and a user). The first chapter highlights the need for a direct consideration of several aspects. Among them are the component of objectivity in scientific representations (i.e., where it can be obtained through the pragmatic approach that van Fraassen proposes); the existence of historicity in the cognitive contents; and the dual orientation of models used in science (i.e., descriptive and prescriptive). It seems clear that, besides the "descriptive" models of basic science, which might be explanatory or predictive models (or both at the same time), there are also "prescriptive" models in applied science, which are related to prediction and prescription.

Within the sphere of the *Philosophical Coordinates* of Part I of the book, Chap. 2 deals with "Scientific Activity as an Interpretative Practice. Empiricism,

Constructivism and Pragmatism." Inmaculada Perdomo stresses philosophy of science as an interpretation of scientific activity, where van Fraassen criticizes realist as well as instrumentalist viewpoints on science in order to build up his interpretative version of such activity. On the one hand, she sees complexity regarding theoretical construction and data generation; and, on the other hand, Perdomo recognizes a central role of the subjects as interpreters of the scientific representations. Thus, she appreciates a family resemblance between some brands of empiricism (constructivist and structuralist) and central tenets of American pragmatism. They are related to the role played by the subject who interprets, constructs and uses models in scientific contexts.

Perdomo insists on this pragmatist component of scientific representation. Thus, besides the representation as relation regarding structures, what is needed is a subject (individual or collective, within a context that gives adequate signs and meanings) in order to express the intentionality in such representation. This involves a process of decision-making where values, aims, and criteria can play a crucial role. Thus, scientists are seeking specific purposes in connection with well defined targets. Consequently, Chap. 2 proposes to look forward to a conception of scientific representation where the social and pragmatic components of science are emphasized. This perspective requires paying attention to dynamic aspects that are characteristic of scientific practice.

Chapter 3 moves toward models and their relation to phenomena, looking at what makes a scientific model a successful representation of its target. In "Models and Phenomena: Bas van Fraassen's Empiricist Structuralism," Valeriano Iranzo discusses the issue of what makes a scientific model a successful representation of its target. In his analysis, this topic begins with models that are intended to represent phenomena, where a successful representation requires an isomorphism about the empirical domain. This isomorphism or *embedment* is between the theoretical model and the data model. But in *Scientific Representation* van Fraassen considers that the structures involved in embedment are not *in re*, a position that can be described in terms that phenomena have no structure while models *are* structures.

On the one hand, Iranzo thinks that the quest for reality should be put aside in van Fraassen's approach, insofar as we get structural knowledge but we should not assume that phenomena themselves have a structure. On the other hand, there is a dependence from the agents: a scientific model is a representation of a target according to the use decided by the agents. Following this indexical component, there is nothing in the target that could account for the fact that this model is a representation of this target system. Iranzo considers the user decision cannot be the last word in this issue: he does not see a sharp distinction between what makes a model a successful representation of its target (isomorphism) and what makes a model a representation of this target instead of another one (users' decision). Thus, scientific representation should be guided by epistemic values, and the model should reveal something about the nature of this particular phenomenon that is its target.

Meanwhile, Part II of this book is on *Representation and Models in Science*. These topics are more directly focused here than in the previous philosophical coordinates. This block begins with a contribution of Bas van Fraassen himself as

Chap. 4. His paper "The Criterion of Empirical Grounding in the Sciences" combines historical cases with philosophical reflections. He pays attention to the interplay of theory, model, and measurement. In this regard, he is interested in two central aspects: (i) what counts as measurement? and (ii) what is measured? Due to the relevance of scientific practice, according to his pragmatic approach, van Fraassen makes an examination of measurement criteria in action.

Bas van Fraassen develops his empiricist approach on empirical grounding. It assumes that a scientific theory offers models for the phenomena in its domain (physical, chemical, etc.); these models involve theoretical quantities of various sorts, and a model's structure is the set of relations it imposes on these quantities. Because of the importance of scientific practice, it is important that those quantities be clearly and feasibly related to *measurement procedures*. This can be seen in historical cases. Thus, van Fraassen examines several scientific episodes: (a) Galileo's measurement of the force of the vacuum; (b) Atwood's machine designed to measure Newtonian theoretical quantities; (c) Michelson and Morley on Fresnel's hypothesis for light aberration; (d) and time-of-flight measurement in quantum mechanics. Bas van Fraassen takes into account the tension between logical strength and relevant evidence. He maintains that the demand for *empirical grounding* has a precise formulation following this scrutiny of crucial junctures, where the role of theory in measurement is highlighted.

An issue directly related to these topics on empirical grounding is "evidence." Maria Carla Galavotti addresses the issue of *evidence* in her paper "On Representing Evidence." Chapter 5 enlarges the epistemological and methodological area of analysis, because "evidence" has only recently directly become a subject field for philosophers of science. Meanwhile, in other fields of research, such as those of law (mainly, penal law) and the health sciences (especially medicine), evidence is — and has been — the focus of extensive debate. Moreover, Galavotti stresses that evidence is a multi-disciplinary subject (and the increasing attention of the philosophers of science regarding topics such as clinical trials, which have clear social consequences, is noticeable). (See, for example, Worrall 2006.)

Galavotti carries out an analysis of the use made of evidence in these two fields: law and the health sciences. She comes to the conclusion that philosophers should pay more attention to the notion of "evidence," its representation and its role for the sake of explanation. In addition, evidence has a clear role in the inferential processes leading to prediction. As a further conclusion, she emphasizes the crucial role of the context concerning evidence (e.g., in the statements made by the judges). She maintains that the analysis of evidence points to a context-sensitive approach to the philosophy of science, which she sees as in tune with van Fraassen's approach.

In Part III of this book, devoted to *Models and Reality*, there are discussions in two realms of the philosophical undertaking developed by van Fraassen for years — logic and epistemology — because he pays attention to logic in order to tackle the philosophical problems of science, and he is commonly stressing the relevance of epistemology for the philosophical analysis of issues raised by scientific activity. In Chap. 6, through his paper "Scientific Models and Abduction: The Role of Non Classical Logic," Ángel Nepomuceno emphasizes abduction, an inferential process that brings about explanatory hypotheses in the context of scientific practice. Abduction shows that the fact to be explained is, in itself, logically independent of the initial theory used to give an explanation of the fact. He sees the "inference to the best explanation" (IBE) as a form of abduction, and he recognizes that van Frasseen critizes IBE.

Nepomuceno analyzes quite different scientific practices — paleoanthropology and the studies of native languages — in order to present the role of abduction in different scientific fields. His main aim in the paper is to point out that the inferential pattern of abduction might not be in classical logic but rather in the non classical logics. In this regard, he thinks that the study made by van Fraassen of Thomason's paradox reveals that we should consider non classical forms of inference that can be used in epistemology. Furthermore, Nepomuceno recognizes the dynamic perspective — in addition to the structural perspective — that can be developed by means of non classical logics as inferential pattern of abduction.

Epistemology is the realm highlighted by Stathis Psillos in his "The View *from Within* and the View *from Above*: Looking at van Fraassen's Perrin." In Chap. 7 he maintains that, in his more recent work, van Fraassen calls attention to the need of theories to be empirically grounded. In this regard, a theory is empirically grounded when its basic theoretical magnitudes are amenable to measurement and the various measurements of the values of these magnitudes yield roughly the same result. He has aimed to explain Jean Perrin's work on Brownian motion and the calculation of Avogadro's number not as a victory of atomism but as a systematic attempt to ground atomic theory empirically, without any commitment to its truth.

Stathis Psillos takes issue with van Fraassen's reconstruction of Perrin's work and argues that Perrin's case shows that it was unreasonable to defend the superiority of the molecular theory c. 1912 without defending its likely truth. Psillos draws attention to van Fraassen's way of viewing the relation between theory and measurement "from above" and "from within," and he examines Perrin's work on Brownian motion from both perspectives. He presents the historical background of Perrin's work and articulates the significance of Perrin's model of Brownian motion for the wider acceptance of atomism. He then offers a probabilistic reconstruction of Perrin's argument for the realities of molecules. In doing all this, Psillos notes that Perrin's work was aiming at more than the empirical grounding of the atomic conception of matter and made its high degree of confirmation possible.

Recently van Fraassen has taken the view that scientific instruments are not "windows on the invisible world" but rather "engines of creation" of new observable phenomena that theories have to save. When it comes to microscopes, van Fraassen has proposed that the phenomena thus created are "public hallucinations."[2] Psillos argues that the robust properties of Brownian motion are not explained by the claim that the "images" observed under microscopes were public hallucinations, on a par with the rainbow. He draws a distinction between offering an intrinsic and an extrinsic explanation of the Brownian images seen under the microscope. Psillos argues

---

[2] In this regard, see the section "The microscope's public hallucinations," in van Fraassen (2008), 101–105. In addition, pages 107–109 are also of interest.

that, while an extrinsic explanation would proceed by aiming to answer the question of why scientists see the particular image as opposed to a different one, the intrinsic explanation — the one that was actually pursued by scientists — required thinking of the image seen under the microscope as representing some genuine effect.

Besides the connection to epistemological topics, Part IV offers additional aspects related to epistemology under the label of *Scientific Explanation and Epistemic Values Judgments*. In Chap. 8, devoted to "Explanation as a Pragmatic Virtue: Bas van Fraassen's Model," Margarita Santana stresses two features. Firstly, the complexity regarding the way of theorizing, insofar as any theorization includes underlying theorizations. The analysis of scientific explanation shows that, when we consider several models and connect them, the theorization itself includes other theorizations that are below it. These theorizations can create the ways of understanding such theorization, which means that no explanatory model is neutral from a metaphysical point of view (including van Fraassen's approach).

Secondly, regarding van Fraassen pragmatic approach to scientific explanation, which has been criticized by Wesley Salmon (see Gonzalez 2002) or Philip Kitcher (see Gonzalez 2011), Santana considers that these criticisms reinforce the previous point on theorization. Futhermore, she thinks that van Fraassen's pragmatic approach to scientific explanation is not good enough, because to include or to consider the context of use (without the inclusion of the agents and the audiences) does not make a model for scientific explanation a pragmatic one. In this regard, van Fraassen seems to agree with this criticism.

Concerning "Values, Choices, and Epistemic Stances," Bas van Fraassen has a quite interesting paper. In Chap. 9 he distinguishes three doxastic tasks (mundane assessments, the tasks of evaluation, and philosophical questions), and he considers the role of value judgments. In this regard, van Fraassen looks backwards to the naturalized epistemology,[3] which was proposed by Willard van Orman Quine. This famous conception appears to make epistemology merely descriptive in form, rather than normative. Thus, in striking contrast with the tradition, it appears that Quine's viewpoint leaves no place for value judgment in rational formation and change of opinion or belief.

van Fraassen recognizes that some more recent forms of naturalism in epistemology are more liberal in this respect, but are still mainly focused on instrumental value alone. The role of value judgment as it appears in epistemic and doxastic tasks faced in science, as well as in more common practical pursuits, is re-examined with a focus on philosophical positions characterized as stances rather than dogmas. He considers that the difference between "first-person" expression of value judgments and "third-person" attribution is crucial to the characterization of tasks involved in our epistemic and doxastic life. The conclusion obtained is that such tasks, at every level, involve value judgment, and that epistemology cannot escape involvement with the normative going beyond instrumental value.

Representation and models in science was the topic of a conference in his honor at the University of A Coruña, Ferrol Campus, where Bas van Fraassen delivered two papers. Some details of this event, *Jornadas sobre Representación y modelos en*

---

[3] On naturalistic approaches see Gonzalez 2006, 1–28; especially, 5–9.

la Ciencia,[4] are in the first chapter of this book, where there is also a large amount of bibliographical information on his work and this relevant philosophical issue. The aim of this information as well as of the volume as a whole is clear: this book discusses at length van Fraassen's approach but it seeks to contribute to the solution of the topic of *representation and models in science*. In this regard, my gratitude to Bas van Fraassen and to all who have cooperated with this shared aim of this book, which is focused on a very relevant thinker of today.

Finally, my recognition again to the persons and institutions that have cooperated in the original event of 2011. First, my appreciation to the speakers of the conference, who are the authors of the papers of this volume; and, second, my acknowledgement to the organizations that gave their support: the Spanish Ministry of Science and Innovation (FFI 2011–12459-E), the City Hall of Ferrol, the University of A Coruña, and the Society of Logic, Methodology, and Philosophy of Science in Spain. In addition, let me point out that I am grateful to Jessica Rey and Amanda Guillán for their contribution to the edition of this book.

13 June 2013   Ferrol, A Coruña

References

The bibliography directly related to the topics indicated here can be found in the first chapter: Gonzalez, W. J. (2014). On representation and models in Bas van Fraassen's approach. In W. J. Gonzalez (Ed.), *Bas van Fraassen's approach to representation and models in science* (Synthese Library, pp. 3–37). Dordrecht: Springer; especially, pp. 17–37. The references used for this Prologue are the following:

Gonzalez, W. J. (Ed.) (2002). *Diversidad de la explicación científica*. Barcelona: Ariel.
Gonzalez, W. J. (2006). Novelty and continuity in philosophy and methodology of science. In W. J. Gonzalez & J. Alcolea (Eds.), *Contemporary perspectives in philosophy and methodology of science* (pp. 1–28). A Coruña: Netbiblo.
Gonzalez, W. J. (Ed.) (2011). *Scientific realism and democratic society: The philosophy of Philip Kitcher* (Poznan studies in the philosophy of the sciences and the humanities). Amsterdam: Rodopi.
van Fraassen, B. C. (1980). *The scientific image*. Oxford: Oxford University Press.
van Fraassen, B. C. (2008). *Scientific representation: Paradoxes of perspective*. New York: Oxford University Press.
Worrall, J. (2006). Why randomise? Evidence and ethics in clinical trials. In W. J. Gonzalez & J. Alcolea (Eds.), *Contemporary perspectives in philosophy and methodology of science* (pp. 65–82). A Coruña: Netbiblo.

---

[4]This was the title of the *XVI Conference on Contemporary Philosophy and Methodology of Science*, a series of workshops coordinated by the author of this prologue and editor of the present book.

Part I
Philosophical Coordinates

Chapter 1
On Representation and Models in Bas van Fraassen's Approach

Wenceslao J. Gonzalez

Abstract Professor van Fraassen's characterization of representation and models requires taking into account several aspects. (1) His intellectual trajectory gives us key elements of the philosophical framework of his approach. (2) The philosophical context of representation and models in science involves the consideration of realms of analysis of representation in science as well as the features of representation in science. (3) van Fraassen's approach to scientific representation and models needs the analysis of representation, understood as activity, and the complex role of models that he presents. (4) Bastiaan Cornelis van Fraassen's publications exemplify the diversity of his contributions, many of them related to these topics. This feature is apparent in four kinds of publications: (i) books as author and editor; (ii) articles and chapters; (iii) reviews, critical notices, replies and comments; and (iv) other publications. (5) Publications on Bas van Fraassen's philosophy offer us analyses of his approach to representation and models in science. (6) Other references of this paper show the variety of sources used in this paper.

Keywords Representation • Models • van Fraassen • Science • Activity

Representation is a key topic in philosophy, in general, and in philosophy of science, in particular. This notion can be used to reconstruct the history of modern philosophy (rationalism, empiricism, and Kantism), and it also plays a crucial role in many recent debates on philosophy of science. In addition, representation receives keen attention in some empirical sciences, such as psychology and cognitive science (cf. Dietrich 2007). Bas van Fraassen has offered us a philosophical approach to

I am grateful to Bas van Fraassen for his remarks on this paper.

W.J. Gonzalez (✉)
Faculty of Humanities, Department of Humanities, University of A Coruña,
Dr. Vazquez Cabrera Street, w/n, 15403-Ferrol (A Coruña), Spain
e-mail: wencglez@udc.es

W.J. Gonzalez (ed.), *Bas van Fraassen's Approach to Representation and Models in Science*, Synthese Library 368, DOI 10.1007/978-94-007-7838-2_1,
© Springer Science+Business Media Dordrecht 2014

representation — in a direct connection to the topic of models — within a new brand of empiricism (a contemporary version built up as a constructivist one). His view includes a different framework for analyzing science (a structural conception) from a pragmatic viewpoint.

The attention shifts here to the coordinates of van Fraassen's approach to representation and models in science. In this regard, there are several steps: first, the main aspects of his intellectual trajectory; second, the philosophical context of representation and models in science, which involves taking into account the realms of analysis in their characterization as well as some key features of representation in science; and third, a review of van Fraassen's approach to scientific representation and models, where representation as activity and the role of models are considered. Thereafter, information is offered on his publications: books as author and editor; articles and chapters; reviews, critical notices, replies and comments; and other publications. These are followed by publications on van Fraassen's philosophy and other references used in this paper.

1.1   Intellectual Trajectory of Bas van Fraassen

Bastiaan Cornelis van Fraassen is a well-known philosopher of science, who was born in Goes (The Netherlands) on April 5, 1941. He has made influential contributions to epistemological and methodological discussions at least since his book *The Scientific Image*, a breakthrough published in 1980 (van Fraassen 1980a). This volume made the author co-winner of the Franklin J. Matchette Prize for Philosophical Books in 1982, as well as co-winner of the Imre Lakatos Award in 1986. This book, in addition to other important publications on an empiricist approach to this field, has contributed to a new recognition to Bas van Fraassen: he received the inaugural 2012 Hempel Award, given by the Governing Board of the Philosophy of Science Association, "recognizing lifetime scholarly achievement in the philosophy of science."[1]

Before these public recognitions, Bas van Fraassen obtained the first degree from the University of Alberta (Canada) in 1963, followed by a Masters degree (in 1964) and a PhD (in 1966) from the University of Pittsburgh (USA). For his PhD van Fraassen worked with Adolf Grünbaum, whose views on representation he analyzed a few years ago (cf. van Fraassen 2009d), and he was quite familiar with Nicholas Rescher's philosophical approach, whose theses on explanation and prediction he has considered in recent years (cf. van Fraassen 2009a). So van Fraassen was well aware of the empiricist viewpoint of the former and the pragmatic conception of the latter.

---

[1] The Governing Board of the Philosophy of Science Association, Official announcement on the inaugural 2012 Hempel Award, September 25, 2012. The Award was given at the PSA meeting in San Diego on November 17, 2012.

These philosophical features — empiricism and pragmatism — appeared later on in van Fraassen but in his own way, because they are embedded within a new philosophical approach to science. As a matter of fact, he offered first a program of "constructive empiricism,"[2] and thereafter, under more reflection, he delivered a "structural empiricism" (cf. van Fraassen 2008a, vii–viii and 237–239). These two philosophical options had clear targets: they have been proposed as explicit alternatives to scientific realism, in general (the former) and to structural realism, in particular (the latter). Although they are interconnected, the initial proposal went clearly in the first anti-realist direction,[3] whereas the subsequent vision moves specifically in the second way.

Within this philosophical trajectory, his book *Scientific Representation: Paradoxes of Perspective* (2008a) offers a clear alternative to structural realism. It is the volume that directly focuses on the topics discussed in this publication. But between his publication of 2008 on scientific representation and the very influential book *The Scientific Image*, published 28 year earlier, there are also several volumes that have raised particular attention among philosophers of science: *Laws and Symmetry* (1989a), *Quantum Mechanics: An Empiricist View* (1991a) and *The Empirical Stance* (2002a). All of them — as well as many papers — are steps in his project of an empiricist version of structuralism.

His invitation to the University of A Coruña (Spain) was also on this route. The main aim was to develop new aspects of his philosophical project to be discussed within a Workshop on his conception, where the focus was on representation and models in science. Bas van Fraassen moves in this novel direction when he presented two papers at the Ferrol Campus on March 11 and 12, 2011. The first text was "Modeling and Measurement: The Criterion of Empirical Grounding," and second one was "The Self, from a Logical Point of View." The former paper is connected with van Fraassen's first contribution to the present volume: "The Criterion of Empirical Grounding in the Sciences," whereas his second contribution to this volume — "Values, Choices, and Epistemic Stances" — develops new ideas regarding his philosophical approach.

Both papers are new steps in his intellectual journey that develops a new empiricism. In his philosophical conception, van Fraassen "advocates a semantic approach to scientific theories and, on that basis, urges skepticism regarding laws of nature, anti-realism regarding unobservables, and pragmatism regarding explanation."[4] The relevance of his contributions to this academic field of the philosophy of science was already recognized when he served as President of the Philosophy of Science Association (1990–1992).

---

[2] "The aim of this book is to develop a constructive alternative to scientific realism," van Fraassen (1980a), vii.

[3] *Science aims to give us theories which are empirically adequate; and acceptance of a theory involves as belief only that it is empirically adequate.* This is the statement of the anti-realist position I advocate; I shall call it *constructive empiricism*," van Fraassen (1980a), 12 (italics are from the original).

[4] The Governing Board of the Philosophy of Science Association, Official announcement on the inaugural 2012 Hempel Award, September 25, 2012.

This Distinguished Professor of Philosophy at San Francisco State University was for 26 years Professor of Philosophy at Princeton University (1982–2008). Before that he taught at the Universities of Yale, Toronto and Southern California. During these decades, besides his research in philosophy of science, in general, and philosophy of physics, in particular, van Fraassen has also developed very important work on philosophy of probability and philosophy of logic. This runs commonly in parallel with his philosophical research on science. Furthermore, as editor of the *Journal of Philosophical Logic* and co-editor of the *Journal of Symbolic Logic*, he has also made a very important contribution. A detailed report of his publications in this area — as well as in the other philosophical branches — can be seen in the bibliography offered below.

1.2 The Philosophical Context of Representation and Models in Science

General philosophy of science has paid a particular attention to representation and models from various angles. Frequently, these play a key role in the analysis of scientific theories. The trend has been particularly noticeable in recent decades. A crucial question is 'how should they be conceived?' Bas van Fraassen has worked hard on this task, and has done this from different angles (logical, epistemological, etc.). But before his position is analyzed, it seems particularly relevant to propose some central components of the contemporary philosophical context of representation and models in science. This setting involves taking into account a relevant characterization of science to be used as a focus for analyzing scientific representation and models nowadays.

1.2.1 Realms of Analysis of Representation in Science

Science is a complex reality that condenses a trajectory of centuries and one that is open to improvement in the future. Thus, the *characteristics of a science* are not simple, but I think that they can be enumerated basically around several elements:[5] (a) science possesses ordinarily a specific language (with terms whose sense and reference are commonly precise); (b) science is articulated in scientific theories with a well patterned internal structure (at least in the most developed theories), which is nevertheless open to later changes (Worrall 2001); (c) science is a qualified knowledge (with more rigor, in principle, than any other human knowledge); (d) science consists of an activity that follows some methods (normally they are deductive, although many authors accept also inductive methods) and it appears as a dynamic activity (of a self-corrective kind).

---

[5] These elements are discussed in Gonzalez 2005, 3–49; especially, 10–11.

In addition to these characteristics of a science, there are other elements that have been emphasized in recent decades: (e) the reality of science comes from social action, and it is an activity whose nature is different from other human activities in its assumptions, contents and limits; (f) science has aims — where cognitive ones are particularly important — for guiding, under the influence of values, its endeavor of researching (in the formal sphere and in the empirical area); and (g) science can have ethical evaluations insofar as it is a free human activity, where certain values might be related to the process itself of research (honesty, originality, reliability …) and some values can be connected with other activities of human life (social, cultural …).

Following these elements, we can find several aspects to be analyzed in science: semantic, logical, epistemological, methodological, ontological, axiological and ethical. To some extent, all of them might be considered regarding scientific representation and models. (i) Semantics of science deals with the language of representation. This requires revising notions such as "resemblance" as well as subjective and objective perspectives related to debates on *Vorstellung* and *Darstellung*. (ii) Logic of science takes into account the "structure" that, within the sphere of scientific theories, needs to be examined in order to see the configuration of a "representation" and its role in models. (iii) Epistemology is focused on the cognitive content involved in a representation. In this regard, the knowledge related to a representation might be subjective, objective or intersubjective. (iv) Methodology of science discusses the progress that might occur in the scientific representation and why it might happen. In this domain there is a clear difference between descriptive models and prescriptive models.[6]

Besides these four "traditional" aspects of philosophy of science, there are another three features to be considered in scientific representation and models. (v) Ontology of science needs to discuss the status of representation as such (i.e., a *Bild* or "mirror" of extramental reality, a social construction, etc.) and the dynamic trait of scientific representation (either in terms of "process," "evolution" or "historicity").[7] (vi) Axiology of research can offer the "internal" and "external" values around scientific representation.[8] There might be values related to the content of the representation (reliability, similitude, etc.) and values accompanying the user and the contextual setting involved in a scientific representation (social, cultural, historical, etc.). (vii) Ethics of science can have a role here insofar as ethical values (endogenous and exogenous) can be connected to epistemological and methodological issues. In this regard, a reliable representation can have also an ethical value.

Most of these seven aspects of science can be detected promptly in van Fraassen's approach to scientific representation and models. He examines some of these realms of analysis in an explicit way, such as the semantic, logical, epistemological and

---

[6] This difference is related to the distinction between basic science and applied science. Cf. Niiniluoto (1993, 1995). This feature is relevant for scientific prediction, see Gonzalez (2006b).

[7] On these three characterizations of scientific change — "process," "evolution" or "historicity" — see Gonzalez (2011).

[8] A broader discussion on the role of values in science is in Gonzalez (2013a).

methodological elements. Other aspects — such as the ontological and axiological elements — receive attention insofar as these facets are connected with those traits that he emphasizes. Thereafter, the ethical factors might be inferred from his writings, because they are rather implicit than explicit in his texts.

Semantics has a role in van Fraassen's approach to representation, which is related to models: "although empiricists in the twentieth century went overboard when they concentrated on linguistic representation, with their syntax (vocabulary and grammar)-oriented view of theories, it is still true that much that was pertinent to representation came along in discussion of language."[9] In this regard, from the beginning of his philosophical career, his perspective on language was according to pragmatics: "we construe the demand for an interpretation of semantic concepts as answerable by the exhibition of a clear pragmatic counterpart" (van Fraassen 1967a, 167).

Pragmatics is then the key for understanding scientific language in van Fraassen, and the pragmatist stance (conceived as the emphasis on human activity) is a central bearing of his philosophical analysis of science. This involves taking into account relations of language(s) to the user and to the context of use. Moreover, he recognizes this, in his book *The Scientific Image*, in which pragmatics touches relevantly on several central issues in philosophy of science (cf. van Fraassen 1980a, ch. 4, sect. 4; ch. 5, sect. 4; and ch. 6, sect. 5). This approach later is reinforced in his volume *Scientific Representation*.

Besides the pragmatic posture on language (and the pragmatist stance), the empiricist structuralist view of science that he develops in *Scientific Representation* emphasizes what I have called here "logic of science." De facto, what van Fraassen offer us in many ways is a *logico-epistemological* view of science: "what scientific theories give us for representing the phenomena are models; models are mathematical structures; mathematical structures are not distinguishable beyond isomorphism; therefore, scientific representation of phenomena does not go beyond representation of their structure" (van Fraassen 2011a, 439).

According to this logico-epistemological view of science, there is in van Fraassen particular attention to logical relations and cognitive abilities. In this regard, Harold I. Brown maintains that, despite the emphasis he puts on the realm of pragmatics for the acceptance and rejection of theories, still "van Fraassen wrestles with essentially the same question that [Karl] Popper does: how can we compare theories — which are abstract entities — with concrete objects in nature" (Brown 2011, 382.) His interest is in how an abstract entity, such as a mathematical structure, can represent something that is not abstract (e.g., objects of nature) (cf. van Fraassen 2008a, 240).

Methodologically, van Fraassen is clearly constructivist on the processes of representation instead of being realist or naturalist. His constructivism regarding representation appears already in *The Scientific Image*, where representation is still mainly diadic (i.e., in a theory a family of *structures* are related to observable *phenomena*) (cf. van Fraassen 1980a, 64). Thereafter, this constructivism became

---

[9] van Fraassen (2011a), 434. Available online, doi: 10.1007/s11016-010-9465-5. Accessed 29 Oct 2012.

increasingly pragmatic in *Scientific Representation*, because the role of agent (the user of the representation) is emphasized and, consequently, the context or setting of the activity of representation (cf. van Fraassen 1980a, 23, 28, and 189).

Nevertheless, here there is again the issue of the logical component insofar as "van Fraassen, like Popper before him, assumes that confirmation and disconfirmation relations are logical relations and thus holds only among abstract items. This raises a problem about how experience, for Popper, and observables, for van Fraassen, enter into epistemic evaluations. Each philosopher offers a drastic proposal: Popper holds that basic statements are accepted by convention; van Fraassen introduces his 'pragmatic tautology'" (Brown 2011, 381).

Meanwhile, epistemology in his approach has more weight than ontology of science. This is particularly clear in his comparison with James Ladyman's approach (cf. Ladyman 2000; and Ladyman 2004). For van Fraassen, there is a crucial difference between that structural realism and his empiricist structuralism:[10] "Structuralism, in the sense of what Ladyman dubbed 'ontic structural realism,' is a *view of what the world is like*, and only derivatively, a view of how science is to be understood. Empiricist structuralism, on the other hand, is a *view of science*, with no implications for what the world is like" (van Fraassen 2011a, 438). The first case — structural realism — is an explicit ontological conception, whereas the second option is an epistemological stance that involves representation as activity and product, where there is no clear-cut implications regarding extramental world.

Axiology of research in van Fraassen's philosophy of science is mainly focused on cognitive values and epistemic utilities, as is the case in his second paper in this volume: "Values, Choices, and Epistemic Stances" (van Fraassen 2014b). In this regard, he highlights as historically important the shift "*from* a focus on truth *to* a focus on representation: to present a theory is to present a family of models, as candidates for *representation* of the phenomena" (van Fraassen 2011a, 435). Thus, the key values in science are not those related to truth, but rather those around representation.

Ethical features might be underneath van Fraassen's approach to representation insofar as he conceives representation as an activity and a product (cf. van Fraassen 2011a, 433–442; especially, 433). He considers that "representation itself (the activity) is *intentional*, both in Brentano's sense and in the common sense of the term." [11] In his view, the role of user of the representation as well as the context of use of the representation are emphasized, which involves *de facto* an intended-use of representation. For him, the intentional use is relevant when "it comes to understanding any form of representation" (van Fraassen 2011a, 437). According to his pragmatic viewpoint, the human activity of representing, which is oriented towards aims chosen by a user, might have an implicit ethical component insofar as representing is a free human act.[12]

---

[10] "Structural realism" appeared explicitly with John Worrall (cf. Worrall, 1989a) and has flourished as a new brand of scientific realism, which involves now a large variety of options.

[11] van Fraassen (2008a), 28. "Literally, 'intentional' refers to intention, but we take it broadly to include purpose, goal, role, and function," (2008a), 181.

[12] See, for example, his remarks on misrepresentation: van Fraassen (2008a), 13–15.

1.2.2   Features of Representation in Science

Commonly, philosophy of science has used "representation" as a key factor to characterize scientific concepts. This is the case, for example, of Paul Thagard when he analyzes "conceptual revolutions," a task that he does from the point of view of concepts as mental representations.[13] But the characterization of the notion of "concept" from the idea of "mental representation" is clearly insufficient,[14] because there are more elements involved in concepts, which include components related to historicity (cf. Gonzalez 2011, 51–55). Before him many authors have thought of *concepts* as a kind of human construction that is linked to representations regarding real or possible phenomena.

Prima facie, the notion of "representation" involves the component of otherness (i.e., alterity) and, in principle, concerns to an intellectual presentation or content of something that might be outside the mind (i.e., an *extramental object or process*). Besides the ontological ingredient of alterity, there is initially a twofold possibility for a representation from the epistemological point of view: a subjective sense (*Vorstellung*) and an objective one (*Darstellung*). In addition to the ontological factor and the epistemological options regarding the content of the representation, there is an agent or user of the representation as well as a public context or environmental setting.

On the one hand, the intellectual presentation of something extramental — actual or possible — can lead to the characterization of "representation" in a subjective sense (*Vorstellung*), when a human agent knows something (a phenomenon or event) and the content is somehow particular or specific of an individual in a given setting (or even for a period of his or her life). On the other hand, the "representation" might be seen in objective terms (*Darstellung*), i.e., as something that it is not the reality in itself (*an sich*), but that possesses public character and can be reached by other minds[15] (e.g., a scientific concept). In this second case, a representation is neither genuinely subjective or private, properly speaking, nor merely intersubjective (a representation of a group or a society).

Subjective representation varies from an individual to another individual, and from one moment of life of an individual to other moment of his or her life. Thus, there are variations regarding the same phenomena or events, not only among individuals but also within an individual in different periods of his or her life (as was the case in Charles Darwin regarding the relations between biological species). Objective representation is when the intellectual presentation of the phenomena or events grasps actual proprieties of the reality represented. This objective representation can appear in a concept (e.g., a mature scientific concept), and it can be understood by different

---

[13] Cf. Thagard (1992), ch. 2, 13–33. "I shall treat concepts and propositions as mental representations (…). In my usage, concepts are mental structures representing what words represent, and propositions are mental structures representing what sentences represent," Thagard (1992), 21.

[14] With his empiricist approach, he goes far beyond this position when he maintains "'mental representation' is an oxymoron," van Fraassen (2008a), 345, note 1.

[15] On the representation as *Vorstellung* and *Darstellung*, cf. Gonzalez (1986), 37–38.

individuals of diverse historical periods, as is the case with scientific concepts proposed by Newton, Darwin or Einstein.

Undoubtedly, it is clear that "representation is a *relational* notion" (van Fraassen 2008a, 26). But this relation of representation of something might be subjective or objective (or even intersubjective). In addition to the role of the agent, van Fraassen agrees that there is a contextual factor: "the very same object or shape can be used to represent different things in different contexts, and in other contexts not represent at all" (van Fraassen 2008a, 27). Moreover, we can think that "a representation trades for its success on *selected* resemblances that deemed *relevant* for the *user* in a certain *context*" (Ghins 2010, 526). Even so, when the representer represents what is represented, is *some* resemblance a *necessary* condition for successful representation?

Michel Ghins thinks such a condition is not needed and, at the same time, he maintains that "the broader notion of *structural similarity* does provide a necessary condition for representation." (Ghins 2010, 526.) At least in science, it seems to me that structural similarity in the relation of representation is needed between two things for one to be able to represent the other. Maybe something else should be available for an actual representation: a content that accompanies the structure and makes the represented — objects or processes — identifiable. This is the idea of some kind of resemblance, which does not need to be a *Bild* or picture of the phenomenon or event. At least, this seems to be the case when we have an empirical representation, because mathematical representation works, due to its abstractiviness, on a different epistemological level.

Certainly, we should emphasize that, when there is a philosophical discussion on how to conceive "representation," two main possibilities appear frequently as general approaches. On the one hand, there is the idea of *representation of*, where a key notion is "resemblance" (i.e., a resemblance between the content of the representation and the reality — object or process — considered). On the other hand, there is the characterization in terms of *representation for*, where the support is based on the notions of "use" and "practice." The first view can receive the endorsement of scientific realists of several sorts;[16] whereas the second perspective is clearly stressed by Bas van Fraassen. He states unequivocally that "*there is no representation except in the sense that some things are used, made, or taken, to represent some things as thus or so*" (van Fraassen 2008a, 23).

If we think of a representation as connected to scientific research, then a distinction should be made between "descriptive representation" and "prescriptive (or 'normative') representation." In the first case, the representation is made in basic science, where the main aim is related to the enlargement of reliable knowledge in the areas of explanation and prediction. Meanwhile in the case of applied science the main goal is the resolution of concrete problems, which involves the use of prediction and prescription.[17] Thus, representation of past or present phenomena

---

[16] On contemporary perspectives on realism in science, see Gonzalez 1993; and Gonzalez, W. J., "Novelty and continuity in philosophy and methodology of science," in Gonzalez 2006a, 1–28; especially, 11–16.

[17] On the distinction between "basic science" and "applied science," see the papers of Ilkka Niiniluoto quoted in note 6. In addition, regarding the connected topics see Gonzalez (1998).

observed or experimented might be different from the representation of phenomena that might come in the future. Therefore, the former might involve less creative representations than the latter, because science related with the future — at least in the case of the sciences of design — requires the constant presence of creativity (cf. Gonzalez 2008).

1.3   van Fraassen's Approach to Scientific Representation and Models

van Fraassen's conception is usually focused on basic science,[18] which is commonly connected to "descriptive representation," and he "hesitates to use such terms as 'applied science'" (van Fraassen 2008a, 360, note 38). His approach to representation is clearly pragmatic and explicitly critical of realist views on representation. In his characterization of representation there is an interplay of at least three elements (cf. Barrett 2009, 635). (i) The *structure* of the representation that is related to the representational item or piece (i.e., what can be found in theoretical models, data models, measurement outcomes, etc.). (ii) The reality itself that plays the role of the *target* or aim of the representation (i.e., the phenomenon or object that has otherness regarding the structure of the representation). (iii) The *researcher* that, within the sphere of a practice, uses such structure related to the reality considered (i.e., the person or individual that conceives the representation thinking of the use of it in connection with a practice).

All in all, there is a *context* (or contexts) to be considered, which might be seen as a fourth element in his approach (cf. van Fraassen 2008a, 28–29). If we analyze this issue in general terms, this is commonly the case: the contexts (historical, social, cultural, economic, etc.) do matter for scientific research.[19] But van Fraassen's view on pragmatics — as happens with the characterization of scientific explanation — emphasizes context regarding the individual doing science in a specific setting, "the sensitivity to *contextual* factors that related interests, concerns, and values. In this respect I gladly admit to working in Rescher's shadow, given how much he has emphasized and advocated a pragmatic over-all orientation in philosophy" (van Fraassen 2009a, 343).

1.3.1   *Representation as Activity*

For van Fraassen, representation is a kind of *activity* — a practice — rather than a pure epistemological relation, which distinguishes his position from other

---

[18] He stresses that the basic aim of science "is empirical adequacy," van Fraassen (2008a), 3.

[19] On this issue of the role of contextual factors seen from the perspective of historicity, cf. Gonzalez 2011, 39–62; especially, 40–55.

influential empiricist approaches. For him, "it is in the activity of representation that representations are produced (…) We lose our topic altogether if we attempt to ask, 'what is a representation?' and tacitly take just one or the other aspect into account; for in fact we cannot understand them in isolation" (van Fraassen 2008a, 7. Cf. Brown 2011, 383). Thus, he sees representation at least as a triadic relation instead of a dyadic relation, because — for van Fraassen — a representation R is made regarding a target T by some person P that uses R in order to represent T (cf. Brown 2011, 383). Consequently, representation cannot be a simple *Bild* or picture of reality without a contextual framework, where two main elements are involved — R and T — instead of at least three (R, T, P).

Following this approach of representation as activity, there is a sort of *intentionality* or purposeful component involved insofar as "nothing is a representation unless it has a certain kind of role in use and practice" (van Fraassen 2008a, 189). This component can be considered at two different levels: on the one hand, in the persons or human agents that have the representations at stake; and, on the other, in the context of research that establishes the parameters to be developed in the ongoing scientific research. Both share the idea of "representation for:" some things *are used* to represent something (cf. van Fraassen 2008a, 23).

First, there is a person or human agent who should have the purpose of using R to represent a target T. Here, according to van Fraassen, there is no "selective likeness" to represent reality (cf. van Fraassen 2008a, 7), instead — even in the case of a common representation such a photo — *"what it is an image of* depends on the use, on *what I use it to represent"* (van Fraassen 2008a, 21). Second, the activity of making a representation "needs extra contextual parameters" (van Fraassen 2008a, 347). In this regard, he stresses that "the purpose for which the representation is made or which it is made to serve" can have a role (cf. van Fraassen 2008a, 347). But they also belong to a research process that goes beyond the isolated representation. Thus, it seems to me reasonable to take into account that, in principle, scientific research as human activity is open to universal components, at least within a dominion of phenomena.[20]

By means of this pragmatic approach to representation, in van Fraassen's approach the main subjective aspects are considered, such as the intentionality or purpose of the agent, and some central intersubjective components, such as the contextual factors related to the research made. But a key issue here is whether the *objectivity* of scientific representation can be obtained through the pragmatic approach he proposes. In addition, the problem of descriptive versus prescriptive account should be addressed: is his approach only a descriptive account of how representation is obtained so far,[21] or is it also a prescriptive account of how representation should be made now and in the future?

---

[20] This is the case even though we should be aware of the limits of methodological universalism, cf. Gonzalez (2012).

[21] "Representation of" and "representation as" are considered explicitly avoiding all relation to mental representation: "I will have no truck with mental representation, in any sense," van Fraassen (2008a), 2.

Given his stress on *empirical adequacy*, Bas van Fraassen seems to focus on a descriptive account of scientific representation rather than on a prescriptive account. In this regard, he hesitates to use the expression "applied science," which is the sphere where sciences make prescriptions after making predictions.[22] Commonly, the analyses on his philosophical work go in the first direction. In this line, the ongoing discussion on the "loss of reality objection" of his pragmatic approach to representation raises doubts regarding that his view can guarantee objectivity in the scientific representation (cf. Ghins 2012).

Insofar as scientific representation is — for van Fraassen — mainly an activity, it involves *de facto* an aim, some process(es), and an expected result. The aim is chosen by the agent or user of the representation; the process(es) require(s) some knowledge, both empirical and conceptual (including mathematical knowledge), and the result is a *structure* that is related to a kind of model (cf. van Fraassen 2008a, 309–312). Scientific representation seems to be an individual activity rather than an activity of a group or community, which implicitly assumes a methodological individualism. The relevance of the specific context is clear or even crucial, according to his view on *indexicality* (cf. van Fraassen 2008a, 59, 181–182, 239, and 259–261). The problem is how then it is possible to reach something genuinely general in science, and how it might have objectivity, which are aspects of relevance for having scientific concepts and for doing research.

If we think of science in historical terms and we accept that *historicity* is another dimension of representation as activity to be considered, then the first question is a comparison: how are the representations involved in a contemporary scientific theory better off or more plausible than the representations of previous scientific theories? The second question is thereafter on structure and content: can there be the case of having a good mathematical structure that might be empirically false from the point of view of the content? Both issues are interwoven, because they require taking into account the change in the *cognitive content* of the concepts used.[23] These improvements in the representations of scientific theories are possible and also having good mathematical structures whose empirical content is not correct, but they should be studied by looking at relevant cases in history of science.[24]

1.3.2   The Role of Models

When van Fraassen's approach to representation and models is reviewed, one of the key aspects is the insistence in the role of *models* in science, which includes different uses of the word "model." His initial view on this topic can be seen in *The Scientific Image*, where the general features of scientific theories involve two components: firstly, "to present a theory is to specify a family of structures, its

---

[22] On the use of prediction and prescription in models, see Simon (1997).

[23] An analysis the change in the cognitive content in science is in Gonzalez (2011), 47–52.

[24] On theses aspects, especially the second one, see Worrall (1989b).

models; and secondly, to specify certain parts of those models (the *empirical substructures*) as candidates for the direct representation of observable phenomena" (van Fraassen 1980a, 64).

Later on, in his book *Scientific Representation*, van Fraassen moves towards an "empiricist structuralism." He maintains that essential to this viewpoint is the core construal of the idea that "all *we know is structure*: (I) Science represents the empirical phenomena as embeddable in certain *abstract structures* (theoretical models). (II) Those abstract structures are describable only up to structural isomorphism" (van Fraassen 2008a, 238). In this conception, representations are connected to structures and models, and they are crucial in his philosophical outlook of science. In this regard, we can think first of a "model" and thereafter how structures and models fit in van Fraassen's scheme of thought on representations.

"Model" is a word that, as van Fraassen recognizes, has some problems: "perhaps it would have been better if the word 'model' had not been adopted by logicians to apply to structures never offered in practice. For undoubtedly, in many contexts, something is called a model only if it is a representation, and the sense in which any solution of an equation is a model of the theory expressed by that equation does not have that meaning. But it is too late to regiment our language so as to correct that, and we will just need to be sensitive to usage in different contexts" (van Fraassen 2008a, 250).

As a matter of fact, van Fraassen uses expressions such as "theoretical models," "data-models," and "surface models." In addition, he also makes it explicit that models are mathematical structures. Then he asks: "But in what sense is it true that *models are mathematical structures*? Only in the same sense that paintings are bits of canvas or wood with paint on them!" (van Fraassen 2011a, 439). This remark visualizes his interest in connecting the cases of science and art, but it seems to me that is not particularly helpful for the purpose of philosophical analysis.

Let us try then a different way of philosophical analysis: a configuration of van Fraassen's approach to models taking a bottom-up line. First, there is a distinction between "phenomena" (conceived as observable objects or processes of any sort) and "appearances" (the contents of measurement outcomes) (cf. van Fraassen 2008a, 283). This involves what is actually to be modeled. In his view, empirical adequacy concerns then phenomena, even though the practice is commonly focused on their appearances. This distinction has two principal consequences, pointed out by Paul Dicken: (i) to articulate "structuralism at the level of representation rather than reality," and (ii) a "broader conception of our criteria of adequacy for a successful scientific theory" (Dicken 2011, 919).

Second, "data models" (i.e., data that are relevant for the models at stake) that might be "surface models" (i.e., an "idealization" of the relevant information available). Data models are constructed from data gathered at various moments of the research made. Thus, the *data model* might summarize the relative frequency of a process found by particular measurements, whereas the *surface model* "smoothes" or "idealizes" mathematically the information already available to replace it in favor of a more sophisticated information (e.g., from the relative frequency to a continuous range of variables) (cf. van Fraassen 2008a, 166–167).

Third, "theoretical models" or specific models within a given theory (cf. van Fraassen 2008a, 238, 240, 245–246, and 248–250). These theoretical models are the vehicles for scientific representation. But, according to Jeffrey Barrett, "since theoretical models are abstract structures and mathematical structures are not distinguished beyond isomorphism, how is it possible for theoretical models to represent phenomena at all? ... In short, if one has only an abstract theoretical structure, then one has no empirical content to test" (Barrett 2009, 636). Furthermore, the relations between "theoretical models" and "data models" do not seem clear enough: "how can we explain how a theory represents phenomena by appeal to a relationship between theoretical models and data models when both of these are abstract entities?" (Barrett 2009, 636).

Finally, there is the possibility of purely abstract models (i.e., mathematical structures as such).[25] But again what matters for van Fraassen is the role that the models play in the use and practice of inquirers. This pragmatic area is the sphere where representation has its place. Moreover, there is an explicit *indexical* understanding of representation (cf. van Fraassen 2008a, 59, 181–182, 239, and 259–261), which emphasizes the "local" use of representations. But then, there is another question pointed out by Barrett: "what exactly might it mean for a theory to be empirically adequate when one has recognized the deeply contingent indexical nature of even measurement outcomes as representations?" (Barrett 2009, 635).

1.3.3   Coda

Bas van Fraassen offers us a very important analysis of scientific representation and the role of models. (i) His approach is mainly logico-epistemologic, and it is focused on basic science rather than on applied science. (ii) His structural empiricism is pragmatic regarding language, empiricist on epistemology, constructivist on methodology — oriented to grasping structures in connection with models —, ontologically pragmatist and with emphasis on cognitive values. (iii) His present position is the outcome of an evolution of his thought in favor of an alternative to structural realism, seeking to place representation where realists like to put truth. (iv) Representation is an activity, which includes a triadic combination of a representational structure, a target and an user. They give context a particular relevance, due to indexicality.

Even though he has developed an important conception on the role of models, which leads to relevant aspects being considered, it seems that van Fraassen still has the problem of guaranteeing an actual epistemological content for science. His insistence that scientific knowledge is only knowledge of structure is not good enough for grasping what science is *de facto* and what it should be. Because philosophical analysis needs to consider how science is made now (and was made in the past) but also how science should be improved towards the future, both as basic

---

[25] This seems to be the case in van Fraassen (1980a), 44.

science and as applied science. New steps seem to be needed to grasp objectivity of scientific concepts and better understanding of new sciences, such as the design science (within the sciences of the artificial),[26] where is quite difficult to work without the idea of some mental representations regarding a possible future.

1.4 Bas C. van Fraassen's Publications

To date van Fraassen has published a large number of texts, most of them of an academic character and a few of a different kind. Pursuing a criterion of relevance for the bibliographical information, van Fraassen's publications are organized here on several levels: (a) books as author and editor; (b) articles in journals and chapters of books; (c) reviews of books and critical notices of publications as well as replies and comments to critics of his views; and (d) other publications, devoted to topics different from philosophy.[27]

1.4.1 Books as Author and Editor

van Fraassen, B. C. (1970a). *An introduction to the philosophy of time and spac*e. New York: Random House. Spanish Translation: van Fraassen, B. C. (1978). *Introducción a la Filosofía del tiempo y el espacio* (J.-P. A. Goicoechea, Trans.). Barcelona: Editorial Labor. Second edition, with new preface and postscript, New York: Columbia University Press, 1985.

van Fraassen, B. C. (1971a). *Formal semantics and logic*. New York: Macmillan. Spanish Translation: van Fraassen, B. C. (1987). *Semántica formal y Lógica* (J. A. Robles, Trans.). Mexico: Universidad Nacional Autónoma de Mexico.

van Fraassen, B. C., & Lambert, K. (Eds.) (1972). *Derivation and counterexample*. Encino: Dickenson. Chinese Translation: van Fraassen, B. C., & Lambert, K. (1975). *Zhe xue de luo ji* (Yongxiang Qian, Trans.) Taiwan: Zhi wen chu ban she.

van Fraassen, B. C. (1980a). *The scientific image*. Oxford: Oxford University Press.[28] Italian edition, with new preface, Bologna 1985. Japanese edition, with new preface, Tokyo 1987. Spanish edition, Mexico, 1995. Chinese edition, Shanghai, 2002. Portuguese edition, Sao Paulo, 2006. Greek edition, with new preface, Athens, 2008.

van Fraassen, B. C., & Beltrametti, E. (Eds.) (1981). *Current issues in quantum logic*. New York: Plenum Press.

van Fraassen, B. C. (1984). *Empirismus im XX. Jahrhundert*. Gesamthochsch: Hagen Fernuniversitaet.

Bencivenga, E., Lambert, K., & van Fraassen, B. (1986). *Logic, bivalence and denotation*. Atascadero: Ridgeview Publication.

van Fraassen, B. C. (1989a). *Laws and symmetry*. Oxford: Oxford University Press. French translation and introduction by C. Chevalley (1994). *Lois et symétrie*. Paris: Vrin.

---

[26] A classic of this field is Simon (1996). The design sciences require the constant use of creativity in their concepts. See, for example, Gonzalez (2013b), pp. 293–305.

[27] I am grateful to Jessica Rey for her contribution to this bibliographical information.

[28] This book is co-winner of the Franklin J. Matchette Prize for Philosophical Books, 1982 as well as co-winner of the Imre Lakatos Award for 1986.

van Fraassen, B. C. (1991a). *Quantum mechanics: An empiricist view*. Oxford: Oxford University Press.
van Fraassen, B. C., Spohn, W., & Skyrms, B. (Eds.) (1991). *Existence and explanation*. Dordrecht: Kluwer.
van Fraassen, B. C. (Ed.) (1997). *Topics in the foundation of statistics*. Dordrecht: Kluwer.
van Fraassen, B. C. (2002a). *The empirical stance*. New Haven: Yale University Press.
van Fraassen, B. C., & Beall J. C. (Eds.) (2003). *Possibilities and paradox: An introduction to modal and many-valued logic*. Oxford: Oxford University Press.
van Fraassen, B. C. (2008a). *Scientific representation: Paradoxes of perspective*. New York: Oxford University Press.

1.4.2   Articles and Chapters

van Fraassen, B. C. (1962a). Capek on eternal recurrence. *Journal of Philosophy, 59*, 371–375.
van Fraassen, B. C. (1966a). Singular terms, truth-value gaps and free logic. *Journal of Philosophy, 63*, 481–495.
van Fraassen, B. C. (1966b). The completeness of free logic. *Zeitschrift fur Mathematik Logik und Grundlagen der Mathematik, 12*, 219–234.
van Fraassen, B. C. (1967a). Meaning relations among predicates. *Nous, 1*, 160–179.
van Fraassen, B. C., & Lambert, K. (1967). On free description theory. *Zeitschrift fur Mathematik Logik und Grundlagen der Mathematik, 13*, 225–240.
van Fraassen, B. C. (1968a). Presupposition, implication, and self-reference. *Journal of Philosophy, 65*, 136–152.
van Fraassen, B. C. (1968b). A topological proof of the Lowenheim-Skolem, compactness, and strong completeness theorems for free logic. *Zeitschrift fur Mathematik. Logik und Grundlagen der Mathematik, 14*, 245–254.
van Fraassen, B. C., & Margenau, H. (1968a). Philosophy of science. In R. Klibansky (Ed.), *Contemporary philosophy. La Philosophie contemporaine* (pp. 25–30). Firenze: La Nuova Italia.
van Fraassen, B. C., & Margenau, H. (1968b). Causality. In R. Klibansky (Ed.), *Contemporary philosophy. La Philosophie contemporaine* (pp. 319–328). Firenze: La Nuova Italia.
van Fraassen, B. C. (1969a). Logical structure in Plato's Sophist. *Metaphysics, 22*, 482–498.
van Fraassen, B. C. (1969b). Conventionality in the axiomatic foundations of the special theory of relativity. *Philosophy of Science, 36*, 64–73.
van Fraassen, B. C. (1969c). Meaning relations and modalities. *Nous, 3*, 155–167.
van Fraassen, B. C. (1969d). Presuppositions, supervaluations, and free logic. In K. Lambert (Ed.), *The logical way of doing things* (pp. 67–91). New Haven: Yale University Press.
van Fraassen, B. C. (1969e). Facts and tautological entailments. *Journal of Philosophy, 66*, 477–487.
van Fraassen, B. C. (1969f). Compactness and Lowenheim-Skolem proofs in modal logic. *Logique et Analyse, 12*, 167–178.
van Fraassen, B. C. (1969g). On Massey's explication of Grünbaum's conception of metric. *Philosophy of Science, 36*, 346–353.
van Fraassen, B. C., & Lambert, K. (1970). Meaning relations, possible objects, and possible worlds. In K. Lambert (Ed.), *Philosophical problems in logic* (pp. 1–20). Dordrecht: Reidel.
van Fraassen, B. C. (1970b). On the extension of Beth's semantics of physical theories. *Philosophy of Science, 37*, 325–334.
van Fraassen, B. C. (1970c). Truth and paradoxical consequences. In R. Martin (Ed.), *The paradox of the liar* (pp. 13–23). New Haven: Yale University Press.
van Fraassen, B. C. (1970d). Inference and self-reference. *Synthese, 21*, 425–438.
van Fraassen, B. C. (1972a). Probabilities and the problem of individuation. In S. Luckenbach (Ed.), *Probabilities, problem and paradoxes* (pp. 121–138). Encino: Dickenson.

van Fraassen, B. C. (1972b). A formal approach to the philosophy of science. In R. Colodny (Ed.), *Paradigms and paradoxes: The philosophical challenge of the quantum domain* (pp. 303–366). Pittsburgh: University of Pittsburgh Press.

van Fraassen, B. C. (1972c). The labyrinth of quantum logics. *Boston Studies in the Philosophy of Science, 13*, 224–254.

van Fraassen, B. C. (1972d). Earman on the causal theory of time. *Synthese, 24*, 87–95. In S. Suppes (Ed.) (1973). *Space, time and geometry* (pp. 85–93). Dordrecht: Reidel.

van Fraassen, B. C. (1972e). The logic of conditional obligation. *Journal Philosophical Logic, 1*, 417–438.

van Fraassen, B. C. (1973a). Values and the heart's command. *Journal of Philosophy, 70*, 5–19.

van Fraassen, B. C. (1973b). Extension, intension, and comprehension. In M. Munitz (Ed.), *Logic and ontology* (pp. 101–131). New York: New York University Press.

van Fraassen, B. C. (1973c). Semantic analysis of quantum logic. In C. A. Hooker (Ed.), *Contemporary research in the foundations and philosophy of quantum theory* (pp. 80–113). Dordrecht: Reidel.

van Fraassen, B. C. (1974a). The formal representation of physical quantities. *Boston Studies in the Philosophy of Science, 13*, 196–209.

van Fraassen, B. C. (1974b). The Einstein-Podolski-Rosen paradox. *Synthese, 29*, 291–309.

van Fraassen, B. C. (1974c). Theoretical entities: The five ways. *Philosophia, 4*, 95–109.[29]

van Fraassen, B. C. (1974d). Bressan and Suppes on modality. In K. F. Schaffner & R. S. Cohen (Eds.), *PSA 1972* (pp. 323–330). Dordrecht: Reidel.

van Fraassen, B. C. (1974e). Hidden variables in conditional logic. *Theoria, 40*, 176–190.

van Fraassen, B. C. (1975a). Incompleteness assertion and Belnap connectives. In D. Hockney & W. Harper (Eds.), *Contemporary research in philosophical logic and linguistic semantics* (pp. 43–70). Dordrecht: Reidel.

van Fraassen, B. C. (1975b). Theories and counterfactuals. In H. N. Castañeda (Ed.), *Action, knowledge and reality: Critical studies in honor of Wilfrid Sellars* (pp. 237–263). Indianapolis: Bobbs-Merrill.

van Fraassen, B. C. (1975c). Platonism's Pyrrhic victory. In A. L. Anderson et al. (Eds.), *The logical enterprise* (pp. 39–50). New Haven: Yale University Press.

van Fraassen, B. C. (1975d). Probabilities of conditionals. In W. Harper & C. A. Hooker (Eds.), *Foundations of probability and statistics* (Vol. 1, pp. 261–308). Dordrecht: Reidel.

van Fraassen, B. C. (1975e). Wilfrid Sellars on scientific realism. *Dialogue, 14*, 606–616.

van Fraassen, B. C., & Hooker, C. A. (1975). A semantic analysis of Bohr's philosophy of quantum mechanics. In W. L. Harper & C. A. Hooker (Eds.), *Foundations of probability and statistics* (Vol. 3, pp. 222–241). Dordrecht: Reidel.

van Fraassen, B. C. (1976a). Report on conditionals. *Teorema, 6*, 5–25.

van Fraassen, B. C. (1976b). Representation of conditional probabilities. *Journal of Philosophical Logic, 5*, 417–430.

van Fraassen, B. C. (1976c). To save the phenomena. *Journal of Philosophy, 73*, 623–632.[30]

van Fraassen, B. C. (1977a). The only necessity is verbal necessity. *Journal of Philosophy, 74*, 71–85.

van Fraassen, B. C. (1977b). The pragmatics of explanation. *American Philosophical Quarterly, 14*, 143–150.[31]

van Fraassen, B. C. (1977c). Relative frequencies. *Synthese, 34*, 133–166.

van Fraassen, B. C. (1977d). On the radical incompleteness of the Manifest image (Comments on Sellars). In F. Suppe & P. Asquith (Eds.), *PSA 1976, Philosophy of Science Association* (Vol. 2, pp. 335–343). East Lansing: PSA.

---

[29] This article, excluding pointless changes, is the same as "Gentle polemics," in van Fraassen (1980a), 204–215.

[30] This article is in part the base for the paper by the same name in van Fraassen (1980a), 41–69.

[31] This text is in part the support for the paper by the same name in van Fraassen (1980a), 97–157.

van Fraassen, B. C. (1978a). Essence and existence. In N. Rescher (Ed.), *American philosophical quarterly monograph series: Studies on ontology* 12 (pp. 1–25). Oxford: Basil Blackwell.
van Fraassen, B. C. (1978b). Time, physical and experiential. *Epistemologia, 1*, 323–338.
van Fraassen, B. C. (1979a). Foundations of probability: A modal frequency interpretation. In G. Toraldo diFrancia (Ed.), *Problems in the foundations of physics* (pp. 344–394). Amsterdam: North-Holland.
van Fraassen, B. C., & Leblanc, H. (1979b). On Carnap and Popper probability functions. *Journal of Symbolic Logic, 44*, 369–373.
van Fraassen, B. C. (1979c). Hidden variables and the modal interpretation of quantum mechanics. *Synthese, 42*, 155–165.
van Fraassen, B. C. (1979d). Propositional attitudes in weak pragmatics. *Studia Logica, 38*, 365–374.
van Fraassen, B. C. (1979e). Modality. In H. Kyburg & P. D. Asquith (Eds.), *Current research in philosophy of science: Proceedings of the PSA critical research problems conference* (pp. 282–290). East Lansing: Philosophy of Science Association.
van Fraassen, B. C. (1979f). Russell's philosophical account of probability. In G. W. Roberts (Ed.), *Bertrand Russell memorial volume* (pp. 384–413). London: Allen and Unwin.
van Fraassen, B. C. (1980b). A temporal framework for conditionals and chance. *Philosophical Review, 89*, 91–108.
van Fraassen, B. C. (1980c). A re-examination of Aristotle's philosophy of science. *Dialogue, 19*, 20–45.
van Fraassen, B. C. (1980d). Rational belief and probability kinematics. *Philosophy of Science, 47*, 165–187.
van Fraassen, B. C. (1981a). Assumptions and interpretations of quantum logic. In B. C. van Fraassen & E. Beltrametti (Eds.), *Current issues in quantum logic* (pp. 17–31). New York: Plenum Press.
van Fraassen, B. C. (1981b). A modal interpretation of quantum mechanics. In B. C. van Fraassen & E. Beltrametti (Eds.), *Current issues in quantum logic* (pp. 229–258). New York: Plenum Press.
van Fraassen, B. C. (1981c). Probabilistic semantics objectified, I. *Journal of Philosophical Logic, 10*, 371–394.
van Fraassen, B. C. (1981d). Probabilistic semantics objectified, II. *Journal of Philosophical Logic, 10*, 495–510.
van Fraassen, B. C. (1981e). Essences and laws of nature. In R. Healey (Ed.), *Reduction, time and reality* (pp. 189–200). Cambridge: Cambridge University Press.
van Fraassen, B. C. (1982a). Rational belief and the common cause principle. In R. McLaughlin (Ed.), *What? Where? When? Why? Essays in Honour of Wesley Salmon* (pp. 193–209). Dordrecht: Reidel.
van Fraassen, B. C. (1982b). The Charybdis of realism: Epistemological implications of Bell's inequality. *Synthese, 5*, 25–38.
van Fraassen, B. C. (1982c). Quantification as an act of mind. *Journal of Philosophical Logic, 11*, 343–369.
van Fraassen, B. C. (1982d). Epistemic semantics defended. *Journal of Philosophical Logic, 11*, 463–464.
van Fraassen, B. C. (1983a). Gentlemen's wagers: Relevant logic and probability. *Philosophical Studies, 43*, 47–61.
van Fraassen, B. C. (1983b). Calibration: A frequency justification for personal probability. In R. S. Cohen & L. Laudan (Eds.), *Physics, philosophy, and psychoanalysis* (pp. 295–319). Dordrecht: Reidel.
van Fraassen, B. C. (1983c). Theory confirmation: Tension and conflict. In P. Weingartner & J. Czermark (Eds.), *Epistemology and philosophy of science: Proceedings of the seventh international Wittgenstein symposium* (pp. 319–329). Vienna: Hoelder-Pichler-Tempsky.
van Fraassen, B. C. (1983d). Shafer on conditional probabilities. *Journal of Philosophical Logic, 12*, 467–470.

van Fraassen, B. C. (1983e). Aim and structure of scientific theories. In R. B. Marcus, G. Dorn, & P. Weingartner (Eds.), *International congress of logic, methodology, and philosophy of science VII* (pp. 307–318). Amsterdam: North-Holland.

van Fraassen, B. C. (1984b). Sulla realta' degli enti matematici. In M. Piatelli-Palmerini (Ed.), *Livelli di Realta'* (pp. 90–110). Milano: Feltrinelli. (*Proceedings of the Conference on Levels of Reality*, Florence, 1978).

van Fraassen, B. C. (1984c). On characterizing Popper and Carnap probability functions. In W. L. Harper, H. Leblanc, R. Gumb, & R. Stern (Eds.), *Essays in epistemology and semantics* (pp. 117–139). New York: Haven Publications.

van Fraassen, B. C. (1984d). Theory comparison and relevant evidence. In J. Earman (Ed.), *Testing scientific theories. Minnesota studies in the philosophy of science* (Vol. 10, pp. 27–42). Minneapolis: University of Minnesota Press.

van Fraassen, B. C. (1984e). Glymour on evidence and explanation. In J. Earman (Ed.), *Testing scientific theories. Minnesota studies in the philosophy of science* (Vol. 10, pp. 165–176). Minneapolis: University of Minnesota Press.

van Fraassen, B. C. (1984f). Belief and the will. *Journal of Philosophy, 81*, 235–256.

van Fraassen, B. C. (1984g). The problem of indistinguishable particles. In J. T. Cushing, C. F. Delaney, & G. M. Gutting (Eds.), *Science and reality: Recent work in the philosophy of science* (pp. 153–172). Notre Dame: University of Notre Dame Press.

van Fraassen, B. C., & Hughes, R. I. G. (1985). Symmetry arguments in probability kinematics. In P. Kitcher & P. Asquith (Eds.), *PSA 1984* (Vol. 2, pp. 851–869). East Lansing: Philosophy of Science Association.

van Fraassen, B. C. (1985a). Statistical behaviour of indistinguishable particles: Problems of interpretation. In P. Mittelstaedt & E.-W. Stachow (Eds.), *Recent developments in quantum logic* (pp. 161–187). Mannheim: Bibliographishes Institüt.

van Fraassen, B. C. (1985b). Empiricism in the philosophy of science. In P. M. Churchland & C. A. Hooker (Eds.), *Images of science: Essays on realism and empiricism, with a reply by Bas C. van Fraassen* (pp. 245–308). Chicago: The University of Chicago Press.

van Fraassen, B. C. (1985c). EPR: When is a correlation not a mystery. In P. Lahti & P. Mittelstaedt (Eds.), *Symposium on the foundations of modern physics: 50 years of the Einstein- Podolsky-Rosen experiment* (pp. 113–128). Singapore: World Scientific.

van Fraassen, B. C. (1985d). ¿Qué son las leyes de la Naturaleza? *Diánoia, 31*, 211–262 (Appeared in 1986).

van Fraassen, B. C., Hughes, R. I. G., & Harman, G. (1986). A problem for relative information minimizers in probability kinematics, continued. *British Journal for the Philosophy of Science, 37*, 453–475.

van Fraassen, B. C. (1986a). A demonstration of the Jeffrey conditionalization rule. *Erkenntnis, 24*, 17–24.

van Fraassen, B. C. (1986b). The world we speak of, and the language we live in. In *Philosophy and culture: Proceedings of the XVII-th world congress of philosophy* (pp. 213–221). Montreal: Editions du Beffroi.

van Fraassen, B. C. (1986c). Identity in intensional logic: Subjective semantics. *Versus, 44/45*, 201–219.[32]

van Fraassen, B. C. (1987a). Symmetries in personal probability kinematics. In N. Rescher (Ed.), *Scientific inquiry in philosophical perspective* (pp. 183–224). Lanham: University Press of America.

van Fraassen, B. C. (1987b). Armstrong on laws and probabilities. *Australasian Journal of Philosophy, 65*, 243–260.

van Fraassen, B. C. (1987c). The semantic approach to scientific theories. In N. J. Nersessian (Ed.), *The process of science* (pp. 105–124). Dordrecht: Martinus Nijhoff.

---

[32] Reprinted in Eco, U., Santambrogio, M., and Violi, P. (Eds.) (1988). *Meaning and mental representation* (pp. 201–220). Bloomington, IN: Indiana University Press.

van Fraassen, B. C. (1988a). Simmetrie e cinematica della probabilità personale. In M. C. Galavotti & G. Gambetta (Eds.), *Epistemologica ed economìa* (pp. 21–42). Bologna: CLUEB.

van Fraassen, B. C. (1988b). The problem of old evidence. In D. F. Austin (Ed.), *Philosophical analysis* (pp. 153–165). Dordrecht: Reidel.

van Fraassen, B. C. (1988c). Symmetry arguments in science and metaphysics. In W. Deppert (Ed.), *Exact sciences and their philosophical foundations* (pp. 385–409). Frankfurt: Verlag Peter Lang.

van Fraassen, B. C. (1988d). The peculiar effects of love and desire. In A. Rorty & B. McLaughlin (Eds.), *Perspectives on self-deception* (pp. 123–156). Berkeley: University of California Press.

van Fraassen, B. C. (1988e). Die Pragmatik des Erklaerens. In G. Schurz (Ed.), *Erklaeren und Verstehen* (pp. 31–56). München: Oldenbourg.[33]

van Fraassen, B. C. (1989b). On explanation in physics. In J. Cushing & E. McMullin (Eds.), *The philosophical consequences of quantum mechanics* (pp. 109–113). Notre Dame: University of Notre Dame Press. (Printed as Appendix to a reprint of (1982). The charybdis of realism: Epistemological implications of Bell's inequality. *Synthese, 52*(1), 25–38).

van Fraassen, B. C. (1989c). Probabilities in physics: An empiricist view. In P. Weingartner & G. Schurz (Eds.), *Philosophy of the natural sciences. Proceedings of the XIIIth Wittgenstein conference* (pp. 339–347). Vienna: Kirchberg Holder-Pichler-Tempsky.

van Fraassen, B. C. (1990a). Figures in a probability landscape. In M. Dunn & A. Gupta (Eds.), *Truth or consequences* (pp. 345–356). Dordrecht: Kluwer.[34]

van Fraassen, B. C. (1991b). Time in physical and narrative structure. In J. Bender & D. E. Wellbery (Eds.), *Chronotypes: The construction of time* (pp. 19–37). Stanford: Stanford University Press.

van Fraassen, B. C. (1991c). The modal interpretation of quantum mechanics. In P. Lahti & P. Mittelstaedt (Eds.), *Symposium on the foundations of modern physics 1990* (pp. 440–460). Singapore: World Publishing Co.

van Fraassen, B. C. (1991d). The problem of measurement in quantum mechanics. In P. Lahti & P. Mittelstaedt (Eds.), *Symposium on the foundations of modern physics 1990* (pp. 497–503). Singapore: World Publishing Co.

van Fraassen, B. C. (1991e). On (Ix)(x = Lambert). In W. Spohn, B. C. van Fraassen, & B. Skyrms (Eds.), *Existence and explanation* (pp. 1–18). Dordrecht: Kluwer.

van Fraassen, B. C. (1991f). La Meccanica Quantistica: uno spettro di interpretazioni. *Iride, 7*, 28–50.

van Fraassen, B. C. (1992a). Jeffrey shifts and self-adjoint operators. *Philosophy of Science, 59*, 163–175.

van Fraassen, B. C. (1992b). La credenza e il problema di Ulisse e le Sirene. In A. E. Galeotti (Ed.), *Individui e Istituzioni* (pp. 77–106). Torino: La Rosa Editrice.

van Fraassen, B. C. (1992c). Faire figure dans un monde probabiliste. In D. Laurier & F. Lepage (Eds.), *Essais sur le Langage et l'Intentionalite'* (pp. 307–322). Montreal/Paris: Bellarmin/Vrin.

van Fraassen, B. C. (1992d). Después del fundacionismo: Entre el círculo vicioso y el regreso al infinito. *Diánoia, 38*, 217–240.[35]

van Fraassen, B. C., & Sigman, J. (1993). Interpretation in science and in the arts. In G. Levine (Ed.), *Realism and representation* (pp. 73–99). Madison: University of Wisconsin Press.

van Fraassen, B. C. (1993a). From vicious circle to infinite regress, and back again. *PSA*, Proceedings of the Philosophy of Science Association Conference, November 1992, (Vol. 2, 6–29) Evanston: Northwestern University Press.

van Fraassen, B. C. (1994a). Against transcendental empiricism. In T. J. Stapleton (Ed.), *The question of hermeneutics: Essays in honor of Joseph J. Kockelmans* (pp. 309–336). Dordrecht: Kluwer.

---

[33] Part I is new, Part II is a reprint of part of Ch. 5 of *The scientific image*, 31–89.

[34] There is an Acrobat version with a correction entered on page 349.

[35] van Fraassen, B. C. (1992). After foundationalism: Between vicious circle and infinite regress. *Proceedings of the XI Simposio de Filosofía del Instituto de Investigaciones Filosóficas.* This paper appeared in 1993.

van Fraassen, B. C. (1994b). The world of empiricism. In J. Hilgevoort (Ed.), *Physics and our view of the world* (pp. 114–134). Cambridge: Cambridge University Press.

van Fraassen, B. C. (1994c). Interpretation of QM: Parallels and choices. In L. Accardi (Ed.), *The interpretation of quantum theory: Where do we stand?* (pp. 7–14). Rome: Instituto della Enciclopedia Italiana, (distributed by Fordham University Press).

van Fraassen, B. C. (1994d). Interpretation of science: Science as interpretation. In J. Hilgevoort (Ed.), *Physics and our view of the world* (pp. 169–187). Cambridge: Cambridge University Press.

van Fraassen, B. C. (1995a). Belief and the problem of Ulysses and the sirens. *Philosophical Studies, 77*, 7–37.

van Fraassen, B. C. (1995b). Against naturalized empiricism. In P. Leonardi & M. Santambrogio (Eds.), *On Quine: New essays* (pp. 68–88). Cambridge: Cambridge University Press.

van Fraassen, B. C. (1995c). Fine-grained opinion, probability, and the logic of belief. *Journal of Philosophical Logic, 24*, 349–377.

van Fraassen, B. C. (1995d). Science, probability, and the proposition. In D. Hull, M. Forbes, & R. M. Brown (Eds.), *PSA94* (Vol. 2, pp. 339–348). East Lansing: Philosophy of Science Association.

van Fraassen, B. C. (1995e). 'World' is not a count noun. *Nous, 29*, 139–157.

van Fraassen, B. C. (1995/1996). A philosophical approach to the foundations of science. *Foundations of Science, 1*, 5–9.

van Fraassen, B. C. (1996a). Science, materialism, and false consciousness. In J. Kvanvig (Ed.), *Warrant in contemporary epistemology: Essays in honor of Alvin Plantinga's theory of knowledge* (pp. 149–181). Littlefield: Rowman.

van Fraassen, B. C. (1997b). Structure and perspective: Philosophical perplexity and paradox. In M. L. Dalla Chiara et al. (Eds.), *Logic and scientific methods* (pp. 511–530). Dordrecht: Kluwer.

van Fraassen, B. C. (1997c). Probabilite' conditionelle et certitude. *Dialogue, 36*, 69–90.[36]

van Fraassen, B. C. (1997d). Putnam's paradox: Metaphysical realism revamped and evaded. *Philosophical Perspectives, 11*, 17–42.

van Fraassen, B. C. (1997e). Sola experientia? Feyerabend's refutation of classical empiricism. *Philosophy of Science, 64*, S385–S395.

van Fraassen, B. C. (1998a). Frequency and the myth of probability. In H. Poser & U. Dirks (Eds.), *Hans Reichenbach: Philosophie im Umkreis der Physik* (pp. 55–67). Berlin: Akademie Verlag.

van Fraassen, B. C. (1998b). The agnostic subtly probabilified. *Analysis, 58*, 212–220.

van Fraassen, B. C. (1999a). The manifest image and the scientific image. In D. Aerts (Ed.), *Einstein meets Magritte: The white book—An interdisciplinary reflection* (pp. 29–52). Dordrecht: Kluwer.

van Fraassen, B. C. (1999b). Conditionalization. A new argument for. *Topoi, 18*, 93–96.

van Fraassen, B. C. (2000a). The false hopes of traditional epistemology. *Philosophy and Phenomenological Research, 60*, 253–280.

van Fraassen, B. C. (2000b). The theory of tragedy and of science: Does nature have narrative structure? In D. Sfendoni-Mentzou (Ed.), *Aristotle and contemporary science* (Vol. 1, pp. 31–59). New York: Peter Lang.

van Fraassen, B. C. (2000c). How is scientific revolution/conversion possible? *Proceedings of the American Catholic Philosophical Association, 73*, 63–80.

van Fraassen, B. C. (2000d). La fin de l'empirisme? *Revue Philosophique de Louvain, 98*, 449–479.

van Fraassen, B. C. (2000e). Michel Ghins on the empirical vs. the theoretical. *Foundations of Science, 30*, 1655–1661.

van Fraassen, B. C. (2001a). Constructive empiricism now. *Philosophical Studies, 106*, 151–170.

---

[36] Revised version of van Fraassen, B. C. (1995). Fine-grained opinion, conditional probability, and the logic of belief.

van Fraassen, B. C. (2002b). Literate experience: The [De-, Re-] interpretation of nature. *Versus, 85/86/87*, 331–358.

van Fraassen, B. C. (2002c). What is empiricism, and what could it be? In B. C. van Fraassen, *The empirical stance* (pp. 31–63). New Haven: Yale University Press.

van Fraassen, B. C. (2003a). On McMullin's appreciation of realism concerning the sciences. *Philosophy of Science, 70*, 479–492.

van Fraassen, B. C. & Monton, B. (2003). Constructive empiricism and modal nominalism. *British Journal for the Philosophy of Science, 54*, 405–422.

van Fraassen, B. C. & Ismael, J. (2003). Symmetry as a guide to superfluous theoretical structure. In K. Brading & E. Castellani (Eds.), *Symmetries in physics: Philosophical reflections* (pp. 371–392). Cambridge: Cambridge University Press.

van Fraassen, B. C. (2004a). Transcendence of the ego: The non-existent knight. *Ratio* (new series)*17*, 453–477.

van Fraassen, B. C. (2004b). Science as representation: Flouting the criteria. *Philosophy of Science, 71*, 794–804.

van Fraassen, B. C. (2005a). Conditionalizing on violated Bell's inequalities. *Analysis, 65*, 27–32.

van Fraassen, B. C. (2005b). Appearance versus reality as a scientific problem. *Philosophic Exchange, 35*, 34–67.

van Fraassen, B. C. (2005c). The day of the dolphins: Puzzling over epistemic partnership. In A. Irvine & K. Peacock (Eds.), *Mistakes of reason: Essays in honour of John Woods* (pp. 111–133). Toronto: University of Toronto Press.

van Fraassen, B. C. (2006a). Vague expectation loss. *Philosophical Studies, 127*, 483–491.

van Fraassen, B. C. (2006b). Weyl's paradox: The distance between structure and perspective. In A. Berg-Hildebrand & Ch. Suhm (Eds.), *Bas C. van Fraassen: The fortunes of empiricism* (pp. 13–34). Frankfurt: Ontos Verlag.

van Fraassen, B. C. (2006c). One hundred and fifty years of philosophy. *Topoi, 25*(1–2), 123–127.

van Fraassen, B. C. (2006d). Le Fictionalisme des Entités Inobservables. *Science et Avenir, Hors-Série: Les fictions de la science* (147), juillet/août.

van Fraassen, B. C. (2006e). Structure: Its shadow and substance. *The British Journal for the Philosophy of Science, 57*, 275–307.

van Fraassen, B. C. (2006f). Representation: The problem for structuralism. *Philosophy of Science, 73*, 536–547.

van Fraassen, B. C. (2007a). Structuralism(s) about science: Some common problems. *Proceedings of the Aristotelian Society Supplementary Volume, 81*, 45–61.

van Fraassen, B. C. (2007b). The constitutive a priori. (Review of *The reign of relativity*, by Ryckman). *Metascience, 16*, 407–419.

van Fraassen, B. C. (2007c). Representation and perspective in science. *Principia, 11*, 97–116.

van Fraassen, B. C., & Muller, F. A. (2008a). How to talk about unobservables. *Analysis, 68*, 197–205.

van Fraassen, B. C. (2008b). Sloughs of despond, mountains of joy. *Critical Quarterly, 50*(4), 74–87.

van Fraassen, B. C., & Peschard, I. (2008c). Identity over time: Objectively, subjectively. *Philosophical Quarterly, 58*, 15–35.

van Fraassen, B. C. (2009a). Rescher on explanation and prediction. In R. Almeder (Ed.), *Rescher studies: A collection of essays on the philosophical work of Nicholas Rescher presented to him on the occasion of his 80th birthday* (pp. 339–361). Berlin: Ontos Verlag.

van Fraassen, B. C. (2009b). The perils of Perrin, at the hands of philosophers. *Philosophical Studies, 143*, 5–24.

van Fraassen, B. C. (2009c). Can empiricism leave its realism behind? In M. Bitbol, P. Kerszberg & J. Petitot (Eds.), *Constituting objectivity: Transcendental approaches of modern physics* (pp. 457–480). Berlin: Ontos Verlag.

van Fraassen, B. C. (2009d). The representation of nature in physics: A reflection on Adolf Grünbaum's early writings. In A. Jokic (Ed.), *Philosophy of physics and psychology: Essays in honor of Adolf Grünbaum* (pp. 305–349). Amherst: Prometheus Books.

van Fraassen, B. C. (2010a). Rovelli's world. *Foundations of Physics, 40*, 390–418.

van Fraassen, B. C. (2011a). Author's response. In J. Ladyman, O. Bueno, M. Suárez & B. C. van Fraassen (Eds.), Scientific representation. A long journey from pragmatics to pragmatics. *Metascience, 20*(3), 433–442. Available online, doi: 10.1007/s11016-010-9465-5. Accessed 29 Oct 2012, 17–26.

van Fraassen, B. C. (2011b). Thomason's paradox for belief, and two consequence relations. *Journal of Philosophical Logic, 40*, 15–32.

van Fraassen, B. C. (2011c). What was Perrin's real achievement? In G. Morgan (Ed.), *Philosophy of science matters: The philosophy of Peter Achinstein* (pp. 231–246). Oxford: Oxford University Press.[37]

van Fraassen, B. C. (2011d). Logic and the philosophy of science. *Journal of the Indian Council of Philosophical Research, 27*, 45–66.

van Fraassen, B. C. (2011e). The physics and metaphysics of identity and individuality. *Metascience, 20*, 225–251.

van Fraassen, B. C. (2012a). Otte on hypotheses in science and religion. In K. Clark & M. Rae (Eds.), *Reason, metaphysics, and mind: New essays on the philosophy of Alvin Plantinga* (pp. 99–106). New York: Oxford University Press.[38]

van Fraassen, B. C. (2012b). Explanation through representation, and its limits (La spiegazione attraverso la rappresentazione, e suoi limiti). *Epistemologia, 35*(1), 30–46.

van Fraassen, B. C. (2012c). Modeling and measurement: The criterion of empirical grounding. *Philosophy of Science, 79*(5), 773–784.

van Fraassen, B. C. (2014a). The criterion of empirical grounding in the sciences. In W. J. Gonzalez (Ed.), *Bas van Fraassen's approach to representation and models in science* (pp. 79–100). Springer: Dordrecht.

van Fraassen, B. C. (2014b). Values, choices, and epistemic stances. In W. J. Gonzalez (Ed.), *Bas van Fraassen's approach to representation and models in science* (pp. 189–211). Springer: Dordrecht.

van Fraassen, B. C. (forthcoming). Simulation: How does an empirical question become mathematical? In S. Vaienti (Ed.), *Simulation and networks*. Paris: Hermann.

1.4.3 Reviews, Critical Notices, Replies and Comments

van Fraassen, B. C. (1967b). A note on Bacon's alternative to Russell. *Philosophical Studies, 18*, 47–48.

van Fraassen, B. C. (1968c). Review of *The logical structure of the world*, by R. Carnap. *Philosophy of Science, 35*, 298–299.

van Fraassen, B. C. (1970e). Rejoinder: On a Kantian conception of language. In R. Martin (Ed.), *The paradox of the liar* (pp. 59–66). New Haven: Yale University Press.

van Fraassen, B. C. (1974f). Putnam on the corroboration of theories. In F. Suppe (Ed.), *The structure of scientific theories* (pp. 434–436). Urbana: University of Illinois Press.

van Fraassen, B. C. (1974g). Review of *Leibniz' philosophy of land language*, by Hide Ishiguro. *Dialogue, 13*, 185–189.

van Fraassen, B. C. (1975f). Comments: Lakoff's fuzzy propositional logic. In D. Hockney et al. (Eds.), *Contemporary research in philosophical logic and linguistic semantics* (pp. 273–278). Hollando: Reild.

van Fraassen, B. C. (1975g). Critical notice: Hilary Putnam, philosophy of logic. *Canadian Journal of Philosophy, 4*, 731–743.

---

[37] With reply by P. Achinstein, 294–296.

[38] Response to Richard Otte's "Theory comparison in science and religion," 86–98.

van Fraassen, B. C. (1977e). Review of *Essays in philosophy and its history*, by Wilfred Sellars. *Annals of Science, 34*, 73–74.

van Fraassen, B. C. (1978c). Review of *Personnelle und Statistische Wahrscheinlichkeit*, by Wolfgang Stegmüller. *Philosophy of Science, 45*, 158–163.

van Fraassen, B. C. (1980e). Critical study: Brian Ellis, rational belief systems. *Canadian Journal of Philosophy, 10*, 497–511.

van Fraassen, B. C. (1981f). Critical study: Paul Churchland, scientific realism and the plasticity of mind. *Canadian Journal of Philosophy, 11*, 555–567.

van Fraassen, B. C. (1981g). Discussion: A problem for relative information minimizers in probability kinematics. *British Journal for the Philosophy of Science, 32*, 375–379.

van Fraassen, B. C. (1985e). Salmon on explanation. *Journal of Philosophy, 82*, 639–651.

van Fraassen, B. C. (1985f). On the question of identification of a scientific theory. (Reply to Pérez Ransanz). *Crítica, 17*, 21–25.

van Fraassen, B. C. (1990b, August 10–16). Review of *Pursuit of truth*, by W. V. O. Quine. *Times Literary Supplement*, 4558, p. 853.

van Fraassen, B. C. (1993b). Armstrong, Cartwright, and Earman on Laws and Symmetry. *Phenomenology and Philosophical Research*, 53, 431–444.

van Fraassen, B. C. (1993c). Three-sided scholarship: Comments on the paper of John R. Donahue, S. J. In E. Stump & T. P. Flint (Eds.), *Hermes and Athena: Biblical exegesis and philosophical theology* (pp. 315–325). Notre Dame: University of Notre Dame Press.

van Fraassen, B. C. (1994d). Gideon Rosen on constructive empiricism. *Philosophical Studies, 74*, 179–192.

van Fraassen, B. C. (1995e). Little review of S. Shapin, A social history of truth. *Common Knowledge, 4*, 81–82.

van Fraassen, B. C. (1997f). Comments on Peter Roeper's. The link between probability functions and logical consequence. *Dialogue, 36*, 27–31.

van Fraassen, B. C. (1997g). Elgin on Lewis' Putnam's paradox. *Journal of Philosophy, 94*, 85–93.

van Fraassen, B. C. (1997h). Modal interpretation of repeated measurement: Reply to Leeds and Healey. *Philosophy of Science, 64*, 669–676.

van Fraassen, B. C (with J. Ladyman, I. Douven, L. Horsten) (1997). A defense of van Fraassen's critique of abductive inference: Reply to Psillos. *The Philosophical Quarterly, 47*(188), 305–321.

van Fraassen, B. C. (1998c). Review of *Interpreting the quantum world*, by J. Bub. *Foundations of Physics, 28*, 683–689.

van Fraassen, B. C. (1999c). Response: Haldane on the future of philosophy. *New Blackfriars, 80*, 177–181.

van Fraassen, B. C. (2000f). The Sham victory of abstraction. (Review of *Conquest of abundance*, by P. Feyerabend). *Times Literary Supplement*, 5073, pp. 10–11.

van Fraassen, B. C. (2004c, February 26). Reply to Chakravarrty, Jauernig, and McMullin. *APA Pacific Division* (127–132). Pasadena. Available online in: http://www.princeton.edu/-fraassen/abstract/ReplyAPA-04.pdf. Accessed 14 Jan 2013.

van Fraassen, B. C. (2004d). Reply to Ladyman, Lipton, and Teller. *Philosophical Studies*. Book symposium, 121(2), 171–192.

van Fraassen, B. C. (2006g). Le quasi-réalisme de Schrödinger (Review of *Schrödinger's Philosophy of quantum mechanics*, by M. Bitbol). *Science et Avenir, Hors-Série*, (148), 80.

van Fraassen, B. C. (2006h). Replies. In A. Berg-Hildebrand & Ch. Suhm (Eds.), *Bas C. van Fraassen: The fortunes of empiricism* (pp. 125–171). Frankfurt: Ontos Verlag.

van Fraassen, B. C. (2007d). Reply: From a view of science to a new empiricism. In B. Monton (Ed.), *Images of empiricism: Essays on science and stances, with a reply from Bas van Fraassen* (pp. 337–383). Oxford: Oxford University Press.

van Fraassen, B. C. (2008c). Review of *Philosophical perspectives on infinity*, by Graham Oppy. *International Philosophical Quarterly, 48*(2), 257–258.

van Fraassen, B. C. (2010b). Scientific representation: Paradoxes of perspective (Precis). *Analysis, 70*, 511–514.

van Fraassen, B. C. (2010c). Reply to Contessa, Ghins, and Healey. *Analysis, 70*, 547–556.
van Fraassen, B. C. (2010d). Precis of scientific representation: Paradoxes of perspective. *Philosophical Studies, 150*, 425–428.
van Fraassen, B. C. (2010e). Reply to Belot, Elgin, and Horsten. *Philosophical Studies, 150*, 461–472.
van Fraassen, B. C. (2011f). On stance and rationality (Reply to contributors). *Synthese, 178*, 155–169.

1.4.4 Other Publications

van Fraassen, B. C. (1976d). The next song. *The Fiddlehead* (3), 33–35.
van Fraassen, B. C., & Alcalay, S. (1978, May). *Picture yourself a philosopher*. USC Graphics.
van Fraassen, B. C. (1979g). Walk to a dead end. *Eros*, 40–43.
van Fraassen, B. C. (1979h). St. Xaviera. *Poetic License, 1*(6), 4–9.
van Fraassen, B. C. (1980f). The tower and the shadow. In B. C. van Fraassen (Ed.), *The scientific image* (pp. 132–134). Oxford: Oxford University Press.
van Fraassen, B. C. (1981h). The game. *Corona, 2*, 103–111.
van Fraassen, B. C. (1989d). A parable. In B. C. van Fraassen (Ed.), *Laws and symmetry* (pp. 59–61). Oxford: Oxford University Press.
van Fraassen, B. C. (1991g). La Città invisibile/The invisible city. In L. Mazza (Ed.) (1988), *World cities and the future of the metropoles: International participations.*[39] Milano: Electa. Published in: *Atlante Metropolitano*, Documents 15 (1991), 113–115.

1.5 Publications on Bas van Fraassen's Philosophy

Certainly the publications on van Fraassen's philosophy are increasing as time is passing by, and it is likely that this process might be intensified after the Hempel Award. In this section there is a selection of books and papers that are related to his work from different angles. In this regard, there is a special interest in those topics connected with the issues of this chapter as well as the desire to emphasize the length of philosophical repercussion from the beginning of his career.

Achinstein, P. (Ed.) (2004). *Science rules: A historical introduction to scientific methods*. Baltimore: Johns Hopkins University Press.
Alspector-Kelly, M. (2001). Should the empiricist be a constructive empiricist? *Philosophy of Science, 68*(4), 413–431.
Alspector-Kelly, M. (2004). Seeing the unobservable: van Fraassen and the limits of experience. *Synthese, 140*(3), 331–353.
Alspector-Kelly, M. (2006). Constructive empiricism and epistemic modesty: Response to van Fraassen and Monton. *Erkenntnis, 64*(3), 371–379.
Armstrong, D. M. (1988). Reply to van Fraassen. *Australasian Journal of Philosophy, 66*(2), 224–229.
Bagir, Z. A. (2003). Bas C. van Fraassen: The empirical stance. *Philosophy of Science, 70*(4), 842–844.

---

[39] *Catáleg de l'exposició: International Exhibition of the XVII Triennale.*

Bandyopadhyay, P. S. (1997). On an inconsistency in constructive empiricism. *Philosophy of Science, 64*(3), 511–514.

Barrett, J. A. (2009). Bas C. van Fraassen. Scientific representation: Paradoxes of perspective. *Journal of Philosophy, 106*(11), 634–639.

Belot, G. (2010). Transcendental idealism among the Jersey metaphysicians. Review of *Scientific representation: Paradoxes of perspective*, by Bas C. van Fraassen. *Philosophical Studies, 150*(3), 429–438.

Berg-Hildebrand, A., & Suhm, Ch. (Eds.) (2006). *Bas C. van Fraassen: The fortunes of empiricism.* Frankfurt: Ontos.

Bird, A. (2003). Kuhn, nominalism, and empiricism. *Philosophy of Science, 70*(4), 690–719.

Bourgeois, W. (1987). On rejecting Foss's image of van Fraassen. *Philosophy of Science, 54*(2), 303–308.

Bressan, A. (1972). Replies to van Fraassen's comments: Bressan and Suppes on modality. *PSA: Proceedings of the Biennial Meeting of the Philosophy of Science Association, 1,* 331–334.

Brown, H. I. (2011). van Fraassen meets Popper: Logical relations and cognitive abilities. *Studies in History and Philosophy of Science, 42*(2), 381–385.

Brown, J. R. (1984). Realism and anthropocentrics. *Proceedings of the Biennial Meeting of the Philosophy of Science Association, 1,* 202–210.

Buekens, F. A. I., & Muller, F. A. (2012). Intentionality versus constructive empiricism. *Erkenntnis, 76*(1), 91–100.

Bueno, O. (1999). Empiricism, conservativeness, and quasi-truth. *Philosophy of Science, 66,* S474–S485.

Bueno, O. (2000). Quasi-truth in quasi-set theory. *Synthese, 125*(2), 33–53.

Bueno, O. (2003a). Is it possible to nominalize quantum mechanics? *Philosophy of Science, 70*(5), 1424–1436.

Bueno, O. (2003b). Review of *The empirical stance*, by Bas C. van Fraassen. *Metascience, 12*(3), 360–363.

Bueno, O. (2006). Representation at the nanoscale. *Philosophy of Science, 73*(5), 617–628.

Bueno, O. (2008). Structural realism, scientific change, and partial structures. *Studia Logica: An International Journal for Symbolic Logic, 89*(2), 213–235.

Buzzoni, M. (1997). Erkenntnistheoretische und ontologische Probleme der theoretischen Begriffe. *Journal for General Philosophy of Science, 28*(1), 19–53.

Callebaut, W. (2012). Scientific perspectivism: A philosopher of science's response to the challenge of big data biology. *Philosophy of Biological and Biomedical Science, 43*(1), 69–80.

Cantini, A. (1990). A theory of formal truth arithmetically equivalent to ID1. *The Journal of Symbolic Logic, 55*(1), 244–259.

Cartwright, N. (1974). van Fraassen's modal model of quantum mechanics. *Philosophy of Science, 41*(2), 199–202.

Cartwright, N. (1993). Defence of 'this Worldly' causality: Comments on van Fraassen's laws and symmetry. *Philosophy and Phenomenological Research, 53*(2), 423–429.

Chalmers, A. (1987). Review of *Images of science: Essays on realism and empiricism, with a reply from Bas C. van Fraassen. Australasian Journal of Philosophy, 65,* 216–218.

Chalmers, A. (2011). Drawing philosophical lessons from Perrin's experiments on Brownian motion: A response to van Fraassen. *The British Journal for the Philosophy of Science, 62*(4), 711–732.

Chandler, J., & Rieger, A. (2011). Self-respect regained. *Proceedings of the Aristotelian Society, 111,* 311–318.

Churchill, J. R. (2006). The empirical stance—Bas van Fraassen. *Religious Studies Review, 32*(3), 177.

Churchland P. M., & Hooker C. A. (Eds.) (1985). *Images of science: Essays on realism and empiricism, with a reply by Bas C. van Fraassen.* Chicago: The University of Chicago Press.

Contessa, G. (2009, March, 30). Review of Bas C. van Fraassen, *Scientific representation: Paradoxes of perspectives. Notre Dame philosophical reviews.* Available online in: http://ndpr.nd.edu/news/23964/?id=15665. Accessed 5 Nov 2012.

Contessa, G. (2010). Empiricist structuralism, metaphysical realism, and the bridging problem. *Analysis, 70*(3), 514–524.

Contessa, G. (2011). Scientific models and representation. In S. French & J. Saatsi (Eds.), *The continuum companion to the philosophy of science* (pp. 120–137). New York: Continuum.

Creath, R. (1985). Taking theories seriously. *Synthese, 62*(3), 317–345.

Creath, R. (1988). The pragmatics of observation. *PSA, 1*, 149–153.

Cross, Ch. B. (1991). Explanation and the theory of questions. *Erkenntnis, 34*(2), 237–260.

Cruse, P. (2007). van Fraassen on the nature of empiricism. *Metaphilosophy, 38*(4), 489–508.

Daniels, Ch. B. (1990). Definite descriptions. *Studia Logica, 49*(1), 87–104.

Dicken, P., & Lipton, P. (2006). What can Bas believe? Musgrave and van Fraassen on observability. *Analysis, 66*(291), 226–233.

Dicken, P. (2009). On the syntax and semantics of observability: A reply to Muller and van Fraassen. *Analysis, 69*(1), 38–42.

Dicken, P. (2011). Review of *Scientific representation: Paradoxes of perspective*, by Bas C. van Fraassen. *Mind, 120*(479), 917–921.

Douven, I. (1996). A paradox for empiricism (?). *Philosophy of Science, 63*, S290–S297.

Douven, I. (1999). Inference to the best explanation made coherent. *Philosophy of Science, 66*, S424–S435.

Douven, I. (2009). Review of *Images of empiricism: Essays on science and stances, with a reply from Bas C. van Fraassen*, by Bradley Monton. *Mind, 118*(470), 504–507.

Ducheyne, S. (2012). Scientific representations as limiting cases. *Erkenntnis, 76*(1), 73–89.

Duncan, H., & Lugg, A. (1988). Review of *Images of science: Essays on realism and empiricism, with a reply from Bas C. van Fraassen*. *Canadian Journal of Philosophy, 18*, 795–804.

Elgin, C. Z. (2010). Keeping things in perspective. Review of *Scientific representation: Paradoxes and perspectives*, by Bas C. van Fraassen. *Philosophical Studies: An International Journal for Philosophy in the Analytic Tradition, 150*(3), 439–447.

Evnine, S. J. (2008). The epistemic shape of a person's life. In S. J. Evnine (Ed.), *Epistemic dimensions of personhood* (pp. 108–138). Oxford: Oxford University Press.

Fine, A. (2001). The scientific image twenty years later. *Philosophical Studies, 106*(1–2), 107–122.

Fischer, R. W. (2011). *Scientific representation: Paradoxes of perspective*, by Bas C. van Fraassen. *Heythorp Journal, 52*(2), 301.

Forbes, C. (2009). Review of Bas van Fraassen's *Scientific representation: Paradoxes of perspective*. *Spontaneous Generations: A Journal for the History and Philosophy of Science, 3*(1), 236–238.

Forster, M. R. (1986). Unification and scientific realism revisited. *PSA, 1*, 394–405.

Foss, J. (1984). On accepting van Fraassen's image of science. *Philosophy of Science, 51*(1), 79–92.

Foss, J. (1991). On saving the phenomena and the mice: A reply to Bourgeois concerning van Fraassen's image of science. *Philosophy of Science, 58*(2), 278–287.

French, S. (1995). Bas van Fraassen quantum mechanics: An empiricist approach. *British Journal for the Philosophy of Science, 46*(3), 436.

Frisch, M. (1999). van Fraassen's dissolution of Putnam's model-theoretic argument. *Philosophy of Science, 66*(1), 158–164.

Ganson, D. A. (2001). *The explanationist defense of scientific reason*. New York: Garland.

Garreta, G. (2002). D'un empirisme sans estravens a une science sans loi. L'agnosticisme epistemologique de Bas van Fraassen. *Critique-París*, (661–662), 501–516.

Gauthier, Y. (1971). *An introduction to the philosophy of the time and space*. Par Bas C. van Fraassen. *Dialogue, 10*(1), 199–201.

Gauthier, Y. (1981). *The scientific image*, par Bas C. van Fraassen. *Dialogue, 20*(3), 579–586.

Ghins, M. (1997). Bas van Fraassen: les lois et la symétrie. *Revue Philosophique de Louvain, 95*(4), 738–754.

Ghins, M. (2000). Présentation du professeur Bas van Fraassen. *Revue Philosophique de Louvain, 98*(3), 445–448.

Ghins, M. (2010). Bas van Fraassen on scientific representation. *Analysis, 70*(3), 524–536.

Ghins, M. (2011a). Models, truth and realism: Assessing Bas van Fraassen's views on scientific representation. *Manuscrito, 34*(1), 207–232. Available online in: http://www.scielo.br/scielo.php?pid=S0100-60452011000100010&script=sci_arttext#tx01. Accessed on 5 Nov 2012.

Ghins, M. (2011b). Scientific representation and realism. *Principia, 15*(3), 461–474.

Ghins, M. (2012). Representation and the loss of reality objection. *Epistemologia, 35*, 47–58.

Giere, R. N. (2005). Scientific realism: Old and new problems. *Erkenntnis, 63*(2), 149–165.

Giere, R. N. (2009). Scientific representation and empiricist structuralism. Essay review of Bas C. van Fraassen's *Scientific representation: Paradoxes of perspective. Philosophy of Science, 76*(1), 101–111.

Gonzalez, W. J. (2006a). Novelty and continuity in philosophy and methodology of science. In W. J. Gonzalez & J. Alcolea (Eds.), *Contemporary perspectives in philosophy and methodology of science* (pp. 1–28). A Coruña: Netbiblo.

Gonzalez, W. J. (2010). Recent approaches on observation and experimentation: A philosophical-methodological viewpoint. In W. J. Gonzalez (Ed.), *New methodological perspectives on observation and experimentation in science* (pp. 9–48). A Coruña: Netbiblo.

Gonzalez, W. J. (Ed.) (2011). *Scientific realism and democratic society: The philosophy of Philip Kitcher*. Amsterdam: Rodopi.

Gordon, B. L. (2003). Ontology schmontology? Identity, individuation, and fock space. *Philosophy of Science, 70*(5), 1343–1356.

Green, M. S., & Hitchcock, Ch. R. (1994). Reflections on reflection: van Fraassen on belief. *Synthese, 98*(2), 297–324.

Grimes, Th. R. (1984). An appraisal of van Fraassen's constructive empiricism. *Philosophical Studies, 45*(2), 261–268.

Groisman, B. (2008). The end of sleeping beauty's nightmare. *The British Journal for the Philosophy of Science, 59*(3), 409–416.

Grossman, N. (1974). The ignorance interpretation defended. *Philosophy of Science, 41*(4), 333–344.

Guerrero Pino, G. (2012). Compromisos epistémicos en el enfoque estructuralista de las teorías. *Revista de Filosofía, 37*(1), 7–26.

Halvoroson, H. (2012). What scientific theories could not be? *Philosophy of Science, 79*(2), 183–206.

Hanna, J. F. (1983). Empirical adequacy. *Philosophy of Science, 50*(1), 1–34.

Hanna, J. F. (1984). On the empirical adequacy of composite statistical hypotheses. *PSA, 1*, 73–80.

Hansen, Ph. P., & Levy, E. (1982). Book review: *The scientific image*, by Bas C. van Fraassen. *Philosophy of Science, 49*(2), 290–293.

Hardcastle, V. G. (1994). The images of observables. *The British Journal for the Philosophy of Science, 45*(2), 585–597.

Hardegree, G. M. (1976). The modal interpretation of quantum mechanics. *PSA, 1*, 82–103.

Harker, D. (2011). A likely explanation: IBE as a guide to better (but not more probable) hypotheses. *South African Journal of Philosophy, 30*(4), 16–28.

Harper, W., & Hajek, A. (1997). Full belief and probability: Comments on van Fraassen. *Dialogue, 36*(1), 91–100.

Hild, M. (1998). The coherence argument against conditionalization. *Synthese, 115*(2), 229–258.

Hitchcock, Ch. R. (1992). Causal explanation and scientific realism. *Erkenntnis, 37*(2), 151–178.

Hitchcock, Ch. R. (1996). The role of contrast in causal and explanatory claims. *Synthese, 107*(3), 395–419.

Hoefer, C. (1993). Review of Bas C. van Fraassen's *Quantum mechanics. Mind, 102*(407), 539–542.

Holdsworth, D. G. (1978). A role for categories in the foundations of quantum theory. *Proceedings of the Biennial Meeting of the Philosophy of Science Association, 1*, 257–267.

Horsten, L. (2004). Bas C. van Fraassen, the empirical stance. *International Studies in the Philosophy of Science, 18*(1), 95–97.

Horsten, L. (2010). Having an interpretation. Review of *Scientific representation: Paradoxes of perspective*, by Bas C. van Fraassen. *Philosophical Studies: An International Journal for Philosophy in the Analytical Tradition, 150*(3), 449–459.

Hughes, R. I. G. (1980). Quantum logic and the interpretation of quantum mechanics. *PSA, 1,* 55–67.

Hunter, B. (2003). Bas van Fraassen, the empirical stance. *Philosophy in Review, 23*(Part 6), 419–422.

Ingarden, R. S. (2000). Modal interpretation of quantum mechanics and classical physical theories. *Lecture Notes in Physics Volume, 539,* 32–51.

Inwagen, P. V. (2007). Impotence and collateral damage: One charge in van Fraassen's indictment of analytical metaphysics. *Philosophical Topics, 35*(1–2), 67–82.

Isaac, A. M. C. (2012). Objective similarity and mental representation. *Australasian Journal of Philosophy, 90,* 1–22.

Jaeger, L. (2006). Bas van Fraassen on religion and knowledge: Is there a third way beyond foundationalist illusion and bridled irrationality? *American Catholic Philosophical Quarterly, 80*(4), 581–602.

Jaworski, W. (2009). The logic of how-questions. *Synthese, 166*(1), 133–155.

Johnston, M. (2003). Objectivity refigurated: Pragmatism without verificationism. In J. Haldane & C. Wright (Eds.), *Reality, representation and projection* (pp. 85–130). New York: Oxford University Press.

Kindi, V. (2011). The challenge of scientific revolutions: van Fraassen's and Friedman's responses. *International Studies in the Philosophy of Science, 25*(4), 327–349.

Kitcher, Ph., & Salmon, W. (1987). van Fraassen on explanation. *The Journal of Philosophy, 84*(6), 315–330.

Kitcher, Ph. (2011). Second thoughts. In W. J. Gonzalez (Ed.), *Scientific realism and democratic society: The philosophy of Philip Kitcher* (pp. 353–389). Amsterdam: Rodopi.

Knuuttila, T. (2011). Modelling and representing: An artefactual approach to model-based representation. *Studies in History and Philosophy of Science (Part A), 42*(2), 262–271.

Koperski, J. (2004). Bas C. van Fraassen: The empirical stance. *Faith and Philosophy, 21*(2), 256–259.

Kronz, F. M. (1993). Review of *Quantum mechanics: An empiricist view*, by Bas C. van Fraassen. *Isis, 84*(3), 620–621.

Kuhn, Th. S. (1992). Introduction. *Proceedings of the Biennial Meeting of the Philosophy of Science Association, 2,* 3–5. East Lansing: Philosophy of Science Association, 1993.

Kukla, A. (1994a). Non-empirical theoretical virtues and the argument from underdetermination. *Erkenntnis, 41*(2), 157–170.

Kukla, A. (1994b). Scientific realism, scientific practice, and the natural ontological attitude. *The British Journal for the Philosophy of Science, 45*(4), 955–975.

Kukla, A. (1995). The two antirealisms of Bas van Fraassen. *Studies in History and Philosophy of Science, Part A, 26*(3), 431–454.

Kukla, A. (1996). Antirealist explanations of the success of science. *Philosophy of Science, 63,* S298–S305.

Kvanvig, J. L. (1994). A critique of van Fraassen's voluntaristic epistemology. *Synthese, 98*(2), 325–348.

Ladyman, J. (2000). What's really wrong with constructive empiricism? van Fraassen and the metaphysics of modality. *British Journal for the Philosophy of Science, 51*(4), 837–856.

Ladyman, J. (2004). Constructive empiricism and modal metaphysics: A reply to Monton and van Fraassen. *The British Journal for the Philosophy of Science, 55*(4), 755–765.

Ladyman, J., Bueno, O., Suárez, M., & van Fraassen, B. C. (2011). Scientific representation. A long journey from pragmatics to pragmatics. *Metascience, 20*(3), 417–442. Available online doi: 10.1007/s11016-010-9465-5. Accessed 29 Oct 2012, 1–26.

Ladyman, J., Ross, D., Spurrett, D., & Collier, J. (2007). Scientific realism, constructive empiricism, and structuralism. In J. Ladyman & D. Ross (with David Spurrett and John Collier), *Every thing must go* (pp. 66–130). Oxford: Oxford University Press.

Lagueux, M. (1994). Friedman's 'instrumentalism' and constructive empiricism in economics. *Theory and Decision, 37*(2), 147–174.

Lagueux, M. (2003). *Explanation in social sciences: Hempel, Dray, Salmon and van Fraassen Revisited*. Montréal: Université du Québec à Montréal.

Lamoreaux, L. (2009). Review of *Scientific representation: Paradoxes of perspective*, by Bas C. van Fraassen. *Dialogue—Kingston Ontario and Montreal, 48*(3), 691–694.

Leeds S. and Healey, R. (1996). A note on van Fraassen's modal interpretation of quantum mechanics. *Philosophy of Science, 63*(1), 91–104.

Leroux, J. (1997). Bas C. van Fraassen, Lois et symetrie. *Dialogue, 36*(1), 203–206.

Levi, I. (1981). Direct inference and conformational conditionalization. *Philosophy of Science, 48*(4), 532–552.

Lipton, P. (1991). Discussion: Contrastive explanation, and causal triangulation. *Philosophy of Science, 58*(4), 687–697.

Lipton, P. (2004). Discussion: Epistemic options. *Philosophical Studies: An International Journal for Philosophy in the Analytic Tradition, 121*(2), 147–158.

Longino, H. (2009). Perilous thoughts: Comment on van Fraassen. *Philosophical Studies, 143*(1), 25–32.

Lowe, E. J. (2004). Review of *Possibilities and paradox: An introduction to modal and many-valued logic*, by J. C. Beall and Bas C. van Fraassen. *History and Philosophy of Science, 25*(Part 4), 329.

Lyre, H. (2010). Humean perspectives on structural realism. *The Philosophy of Science in a European Perspective, 1*, 381–397.

Mackinnon, E. (1979). Scientific realism: The new debates. *Philosophy of Science, 46*(4), 501–532.

Mackinnon, E. (1986). Review of *Images of science: Essays on realism and empiricism, with a reply from Bas C. van Fraassen*, by P. M. Churchland & C. A. Hooker. *Isis, 77*(1), 116–117.

Mackintosh, D. (1994). Partial convergence and approximate truth. *The British Journal for the Philosophy of Science, 45*(1), 153–170.

Maclean, D. (2012). Armstrong and van Fraassen on probabilistic laws of nature. *Canadian Journal of Philosophy, 42*(1), 1–14.

Maher, P. (1990). Acceptance without belief. *PSA: Proceedings of the Biennial Meeting of the Philosophy of Science Association, 1*, 381–392.

Maher, P. (1992). Diachronic rationality. *Philosophy of Science, 59*(1), 120–141.

Massey, G. (1974). Review of *An introduction of the philosophy of time and space*, by Bas C. van Fraassen. *Philosophy of Science, 41*(1), 90–92.

Massimi, M. (2007). Saving unobservable phenomena. *The British Journal for the Philosophy of Science, 58*(2), 235–262.

Maxwell, N. (1993a). Induction and scientific realism: Einstein versus van Fraassen. Part one: How to solve the problem of induction. *British Journal for the Philosophy of Science, 44*(1), 61–79.

Maxwell, N. (1993b). Induction and scientific realism: Einstein versus van Fraassen. Part two: Aim-oriented empiricism and scientific essentialism. *The British Journal for the Philosophy of Science, 44*, 81–101.

Maxwell, N. (1993c). Induction and scientific realism: Einstein versus van Fraassen. Part three: Einstein, aim-oriented empiricism and the discovery of special and general relativity. *British Journal for the Philosophy of Science, 44*(2), 275–305.

Maxwell, N. (2009). Muller's critique of the argument for aim-oriented empiricism. *Journal for General Philosophy of Science, 40*(1), 103–114.

McMichael, A. (1985). van Fraassen's instrumentalism. *British Journal for the Philosophy of Science, 36*, 257–272.

McMullin, E. (2003). van Fraassen's unappreciated realism. *Philosophy of Science, 70*(Part 3), 455–478.

Millgram, E. (2006). Bas van Fraassen, the empirical stance. *Philosophical Review, 115*(3), 404–408.

Minogue, B. P. (1984). van Fraassen's semanticism. *PSA, 1*, 115–121.

Minogue, B. P. (1987). Review of *Images of science: Essays on realism and empiricism, with a reply from Bas C. van Fraassen. Zygon, 22*, 257–261.

Mitchell, S. (1988). Constructive empiricism and anti-realism. *PSA, 1*, 174–180.

Mitchell, S. (2010). *Scientific representation: Paradoxes of perspective*, by Bas C. van Fraassen. *Metaphilosophy, 41*(5), 717–722.

Monton, B. J. (1999). van Fraassen and Ruetsche on preparation and measurement. *Philosophy of Science, 66*(3), S82–S91.

Monton, B. (2005). van Fraassen, Bastiaan Cornelis. In J. R. Shook (Ed.) *The dictionary of modern American philosophers* (1st ed., pp. 2472–2476). Bristol: Continuum.

Monton, B. J. (Ed.) (2007). *Images of empiricism: Essays on science and stances, with a reply from Bas C. van Fraassen*. Oxford: Oxford University Press.

Morrison, M. (1990). Theory, intervention, and realism. *Synthese, 82*(1), 1–22.

Morrison, M. (2007). Where have all the theories gone? *Philosophy of Science, 74*(2), 195–228.

Muller, F. A. (2004a). Can a constructive empiricist adopt the concept of observability? *Philosophy of Science, 71*(1), 80–97.

Muller, F. A. (2004b). Erratum: Can a constructive empiricist adopt the concept of observability? *Philosophy of Science, 71*(4), 635–636.

Muller, F. A. (2005). The deep black sea: Observability and modality afloat. *The British Journal for the Philosophy of Science, 56*(1), 61–99.

Muller, F. A. (2008). In defence of constructive empiricism: Maxwell's master argument and aberrant theories. *Journal for General Philosophy of Science, 39*(1), 131–156.

Muller, F. A. (2009). The insidiously enchanted forrest: Essay review of 'Scientific Prediction' by Bas C. van Fraassen. *Studies in History and Philosophy of Moderns Physics, 40*(3) 268–272.

Murad, M. H. S. Bin A. (2012). Models, scientific realism, the intelligibility of nature, and their cultural significance. *Studies in History and Philosophy of Science, 42*(2), 253–261.

Musgrave, A. (1999). Realism versus constructive empiricism. In A. Musgrave (Ed.), *Essays on realism and rationalism* (pp. 106–130). Amsterdam: Rodopi.

Nelson, D. E. (1996). Confirmation, explanation, and logical strength. *The British Journal for the Philosophy of Science, 47*(3), 399–413.

Norris, Ch. (1997). Ontology according to van Fraassen: Some problems with constructive empiricism. *Metaphilosophy, 28*(3), 196–218.

Okasha, S. (2000). van Fraassen's critique of inference to the best explanation. *Studies in History and Philosophy of Science, 31*(4), 691–710.

Okruhlik, K. (2009). Critical notice of Bas C. van Fraassen *Scientific representation: Paradoxes of perspective. Canadian Journal of Philosophy, 39*(4), 671–694.

Padovani, F. (2012). Bas C. van Fraassen: *Scientific representation: Paradoxes of perspective. Science and Education, 21*(8), 1199–1204.

Paris, J. B., & Vencovská, A. (2011). Symmetry's end? *Erkenntnis, 74*(1), 53–67.

Perdomo Reyes, I., & Sánchez Navarro, J. (2003). *Hacia un nuevo Empirismo: La propuesta filosófica de Bas C. van Fraassen*. Madrid: Biblioteca Nueva.

Perdomo Reyes, I. (2011). The characterization of epistemology in Philip Kitcher: A critical reflection from new empiricism. In W. J. Gonzalez (Ed.), *Scientific realism and democratic society: The philosophy of Philip Kitcher* (pp. 113–138). Amsterdam: Rodopi.

Peressini, A. (1999). Confirming mathematical theories: An ontologically agnostic stance. *Synthese, 118*(2), 257–277.

Pincock, C. (2011). Bas C. van Fraassen scientific representation: Paradoxes of perspective. *British Journal for the Philosophy of Science, 62*(3), 677–682.

Pool, M. (2000). How liberating is van Fraassen's voluntarism? *Dialogue, 39*(3), 475–490.

Psillos, S. (1996). On van Fraassen's critique of abductive reasoning. *The Philosophical Quarterly, 46*(182), 31–47.[40]

Psillos, S. (2007). Putting a bridle on irrationality: An appraisal of van Fraassen's new epistemology. In B. Monton (Ed.), *Images of empiricism* (pp. 134–164). Oxford: Oxford University Press.

Psillos, S. (2012). One cannot be just a little bit realist: Putnam and van Fraassen. In R. J. Brown (Ed.), *Philosophy of science: The key thinkers* (pp. 188–212). New York: Continuum.

Reddam, J. P. (1982). van Fraassen on propositional attitudes. *Philosophical Studies, 42*(1) 101–110.

Regt, H. C. D. G. de (2006). To believe in belief. Popper and van Fraassen on scientific realism. *Journal for General Philosophy of Science, 37*(1), 23–40.

Ribeiro, C. (2009). *Electrões inobserváveis e estrelas invisíveis: em torno do problema do realismo em ciência. Bas C. van Fraassen versus Alan Musgrave.* Lisbon: Facultade de Ciências da Universidade de Lisboa.

Richardson, A. W., & Uebel, Th. E. (2005). The epistemic agent in logical positivism. *Proceedings of the Aristotelian Society, Supplementary Volumes, 79*, 73–87 and 89–105.

Richardson, A. (2011). But what then am I, this inexhaustible, unfathomable historical self? Or, upon what ground may one commit empiricism? *Synthese, 178*(1), 143–154.

Ritchie, J. (2012). Styles for Philosophers of Science. *Studies in History and Philosophy of Science, 43*(4), 649–656.

Rorty, R. (2002, July 7). Review of Bas C. van Fraassen, *The empirical stance. Notre dame philosophical reviews.* Available online in: http://ndpr.nd.edu/news/23164-the-empirical-stance/. Accessed 5 Nov 2012.

Rowbottom, D. P. (2009). Images of van Fraassen. *Metascience, 18*(2), 307–312.

Ruetsche, L. (1996). van Fraassen on preparation and measurement. *Philosophy of Science, 63*(3), S338–S346.

Russo, F. (2006). Salmon and van Fraassen on the existence of unobservable entities: A matter of interpretation of probability. *Foundations of Science, 11*(3), 221–247.

Ruttkamp, E. (2002). *A model-theoretic realist interpretation of science,* (Synthese Library, v. 311). Boston: Kluwer.

Sandborg, D. (1998). Mathematical explanation and the theory of why-questions. *The British Journal for the Philosophy of Science, 49*(4), 603–624.

Sankey, H. (1997). van Fraassen's constructive empiricism. *Cogito, 11*(3), 175–182.

Schupbach, J. N. (2007). Must the scientific realist be a rationalist? *Synthese, 154*(2) 329–334.

Schwarz, D. S. (1977). On pragmatic presupposition. *Linguistics and Philosophy, 1*(2) 247–257.

Seager, W. (1988). Scientific anti-realism and the epistemic community. *PSA, 1*, 181–187.

Seager, W. (1995). Ground truth and virtual reality: Hacking vs. van Fraassen. *Philosophy of Science, 62*(3), 459–478.

Sfendoni-Mentzou, D. (2008). Bas van Fraassen's 'Argument from Public Hallucination' and the quest for the real behind representations. *Philosophy of Natural Sciences* (Proceedings of the XXII World Congress of Philosophy), *43*, 199–205.

Sicha, J. F. (1992). Review of *Images of science: Essays on realism and empiricism*, by P. M. Churchland & C. A. Hooker. *Nôus, 26*(4), 519–525.

Simard, J.-C. (2011). Review of *Scientific representation: Paradoxes of perspective*, by Bas C. van Fraassen. *Philosophiques, 38*(2), 615–619.

Simon, S., & Moraes, A. (2007). O empirismo construtivo de Bas C. van Fraassen e o problema do sucesso científico. *Philósophos, 12*(2), 131–169.

Slowik, E. (2012). On structuralism's multiple paths through spacetime theories. *European Journal for Philosophy of Science, 2*(1), 45–66.

---

[40] Translation into Portuguese: Sobre a critica de van Fraassen ao raciocinio abdutivo. *Crítica. Revista de Filosofía*, 6, (2000), 35–62.

Smith, J. M. (1993). Book review: *Laws and symmetry*, by Bas C. van Fraassen. *Philosophy of Science, 60*(4), 661–662.

Sober, E. (1984). Common cause explanation. *Philosophy of Science, 51*(2), 212–241.

Sober, E. (1985). Constructive empiricism and the problem aboutness. *The British Journal for the Philosophy of Science, 36*(1), 11–18.

Spehrs, A. (2010). Comentarios bibliográficos, *Scientific Representation: Paradoxes of Perspectives. Revista iberoamericana de Filosofía, 36*(1), 131–134.

Stairs, A. (1984). Sailing into the Charybdis: van Fraassen on Bell's theorem. *Synthese, 61*(3), 351–359.

Stanford, P. K. (2000). An antirealist explanation of the success of science. *Philosophy of Science, 67*(2), 266–284.

Suárez, M. (2011). Scientific realism, the Galilean strategy, and representation. In W. J. Gonzalez (Ed.), *Scientific realism and democratic society: The philosophy of Philip Kitcher* (pp. 269–292). Amsterdam: Rodopi.

Tang, P. C. L. (1992). Book review: *Laws and symmetry*, by Bas C. van Fraassen. *Isis, 83*(1), 176–177.

Tappenden, P. (2011). Evidence and uncertainty in Everett's multiverse. *The British Journal for the Philosophy of Science, 62*(1), 99–123.

Teller, P. (2001). Whither constructive empiricism? *Philosophical Studies, 106*, 123–150.

Teller, P. (2004). Discussion: What is a stance? *Philosophical Studies: An International Journal for Philosophy in the Analytical Tradition, 121*(2), 159–170.

Temple, D. (1988). Discussion: The contrast theory of why-questions. *Philosophy of Science, 55*(1), 141–151.

Thalos, M. (1999). In favor of being only humean. *Philosophical Studies: An International Journal for Philosophy in the Analytic Tradition, 93*(3), 265–298.

Thomson-Jones, M. (2011). Review of *Scientific representation: Paradoxes of perspective*, by Bas C. van Fraassen. *Australasian Journal of Philosophy, 89*(3), 567–570.

Thornton, M. (1981). Sellar's scientific realism: A reply to van Fraassen. *Dialogue, 20*(1), 79–83.

Tooley, M. (1995). Bas C. van Fraassen *Laws and symmetry*. *British Journal for the Philosophy of Science, 46*(2), 280.

Tregear, M. (2004). Utilising explanatory factors in induction? *The British Journal for the Philosophy of Science, 55*(3), 505–519.

Tregenza, B. (2005). van Fraassen, Bas C. In T. Honderich (Ed.), *The Oxford companion to philosophy* (2nd ed., pp. 896-ff). Oxford: Oxford University Press.

Trout, J. D. (1992). Theory-conjunction and mercenary reliance. *Philosophy of Science, 59*(2), 231–245.

Turney, P. (1990). Embeddability, syntax, and semantics in accounts of scientific theories. *Journal of Philosophical Logic, 19*(4), 429–451.

Vickers, P. (2011). A brief chronology of the philosophy of science. In S. French & J. Saatsi (Eds.), *The continuum companion to the philosophy of science* (pp. 359–380). New York: Continuum.

Vollmer, S. (2000). Two kinds of observation: Why van Fraassen was right to make a distinction, but made the wrong one. *Philosophy of Science, 67*(3), 355–365.

Walden, K. (2005). On taking stances. An interview with Bas van Fraassen. *The Harvard Review of Philosophy, 13*(2), 86–102.

Weisberg, J. (2007). Conditionalization, reflection, and self-knowledge. *Philosophical Studies, 135*(2), 179–197.

Weisberg, J. (2009). Locating IBE in the Bayesian framework. *Synthese, 167*(1), 125–143.

Wessels, L. (1974). Laws and meaning postulates in van Fraassen's view of theories. *PSA: Proceedings of the Biennial Meeting of the Philosophy of Science Association, 1*, 215–234.

Wilson, N. L. (1973). *Formal semantics and logic*, by Bas C. van Fraassen. *Dialogue, 12*(1), 150–151.

Wray, K. B. (2007). A selectionist explanation for the success and failures of science. *Erkenntnis, 67*(1), 81–89.

Wray, K. B. (2010). Selection and predictions success. *Erkenntnis, 72*(3), 365–377.

1.6   Other References of This Paper

Some of the references in this section belong to volumes of the *Gallaecia Series. Studies in Philosophy and Methodology of Science*. This collection deals with central topics in contemporary philosophy of science.[41] Thus, they include conceptions and approaches that are complementary or alternative to Bas van Fraassen's views on representation and models. In addition, there are references of papers mentioned in this chapter that offer helpful insights regarding the topics here discussed.

Dietrich, E. (2007). Representation. In P. Thagard (Ed.), *Philosophy of psychology and cognitive science* (pp. 1–29). Amsterdam: Elsevier.

Gonzalez, W. J. (1986). *La Teoría de la Referencia. Strawson y la Filosofía Analítica*, Salamanca-Murcia: Ediciones Universidad de Salamanca y Publicaciones de la Universidad de Murcia.

Gonzalez, W. J. (1993). El realismo y sus variedades: El debate actual sobre las bases filosóficas de la Ciencia. In A. Carreras (Ed.), *Conocimiento, Ciencia y Realidad* (pp. 11–58). Saragosa: Seminario Interdisciplinar de la Universidad de Zaragoza-Ediciones Mira.

Gonzalez, W. J. (1998). Prediction and prescription in economics: A philosophical and methodological approach. *Theoria, 13*(32), 321–345.

Gonzalez, W. J. (2005). The philosophical approach to science, technology and society. In W. J. Gonzalez (Ed.), *Science, technology and society: A philosophical perspective* (pp. 3–49). A Coruña: Netbiblo.

Gonzalez, W. J. (2006b). Prediction as scientific test of economics. In W. J. Gonzalez & J. Alcolea (Eds.), *Contemporary perspectives in philosophy and methodology of science* (pp. 83–112). A Coruña: Netbiblo.

Gonzalez, W. J. (2008). Rationality and prediction in the sciences of the artificial: Economics as a design science. In M. C. Galavotti, R. Scazzieri & P. Suppes (Eds.), *Reasoning, rationality and probability* (pp. 165–186). Stanford: CSLI Publications.

Gonzalez, W. J. (2011). Conceptual changes and scientific diversity: The role of historicity. In W. J. Gonzalez (Ed.), *Conceptual revolutions: From cognitive science to medicine* (pp. 39–62). A Coruña: Netbiblo.

Gonzalez, W. J. (2012). Methodological universalism in science and its limits: Imperialism versus complexity. In K. Brzechczyn & K. Paprzycka (Eds.), *Thinking about provincialism in thinking* (Poznan studies in the philosophy of the sciences and the humanities, Vol. 100, pp. 155–175). Amsterdam/New York: Rodopi.

Gonzalez, W. J. (2013a). Value ladenness and the value-free ideal in scientific research. In Ch. Lütge (Ed.), *Handbook of the philosophical foundations of business ethics* (pp. 1503–1521). Dordrecht: Springer.

---

[41] As editor of the volumes, I can summarized the titles and the years of publication: *Progreso científico e innovación tecnológica*, 1997; *El Pensamiento de L. Laudan. Relaciones entre Historia de la Ciencia y Filosofía de la Ciencia* 1998; *Ciencia y valores éticos*, 1999; *Problemas filosóficos y metodológicos de la Economía en la Sociedad tecnológica actual*, 2000; *La Filosofía de Imre Lakatos: Evaluación de sus propuestas*, 2001; *Diversidad de la explicación científica*, 2002; *Análisis de Thomas Kuhn: Las revoluciones científicas*, 2004; *Karl Popper: Revisión de su legado*, 2004; *Science, Technology and Society: A Philosophical Perspective*, 2005; *Evolutionism: Present Approaches*, 2008; *Evolucionismo: Darwin y enfoques actuales*, 2009; *New Methodological Perspectives on Observation and Experimentation in Science*, 2010; *Conceptual Revolutions: From Cognitive Science to Medicine*, 2011; *Scientific Realism and Democratic Society: The Philosophy of Philip Kitcher*, 2011; and *Freedom and Determinism: Social Sciences and Natural Sciences*, 2012.

Gonzalez, W. J. (2013b). The sciences of design as sciences of complexity: The dynamic trait. In H. Andersen, D. Dieks, W. J. Gonzalez, Th. Uebel & G. Wheeler (Eds.), *New challenges to philosophy of science* (pp. 293–305). Dordrecht: Springer.

Niiniluoto, I. (1993). The aim and structure of applied research. *Erkenntnis, 38*, 1–21.

Niiniluoto, I. (1995). Approximation in applied science. *Poznan Studies in the Philosophy of the Sciences and the Humanities, 42*, 127–139.

Simon, H. A. (1996). *The sciences of the artificial* (3rd ed.). Cambridge, MA: The MIT Press. (1st ed., 1969; 2nd ed., 1981).

Simon, H. A. (1997). Prediction and prescription in systems modeling. *Operations Research, 38*, 7–14. (Reprinted in: Simon, H. A. (1997). *Models of bounded rationality*. Vol. 3: *Empirically grounded economic reason* (pp. 115–128). Cambridge, MA: The MIT Press.

Thagard, P. (1992). *Conceptual revolutions*. Princeton: Princeton University Press.

Worrall, J. (1989a). Structural realism: The best of both worlds? *Dialectica, 43*, 99–124. (Reprinted in: Papineau, D. (Ed.) (1996). *Scientific realism* (pp. 139–165). Oxford: Oxford University Press).

Worrall, J. (1989b). Fresnel, Poisson and the white spot: The role of successful predictions in the acceptance of scientific theories. In D. Gooding, T. Pinch & S. Schaffer (Eds.), *The uses of experiment* (pp. 135–157). Cambridge: Cambridge University Press.

Worrall, J. (2001). Programas de investigación y heurística positiva: Avance respecto de Lakatos. In W. J. Gonzalez (Ed.), *La Filosofía de Imre Lakatos: Evaluación de sus propuestas* (pp. 247–268). Madrid: UNED.

Chapter 2
Scientific Activity as an Interpretative Practice. Empiricism, Constructivism and Pragmatism

Inmaculada Perdomo

Abstract Since the publication of *The Scientific Image* and earlier works Bas C. van Fraassen has defended his constructive empiricism as the most appropriate philosophical interpretation of scientific activity in critical open dialogue with realisms (both old and new) and instrumentalisms. A new impetus was added to the debate by the publication of his most recent book, *Scientific Representation*, in which he qualifies some of his basic suppositions and proposes a new name for his empiricism: *empiricist structuralism*. In this paper I argue in line with his thesis that if philosophy of science aims to offer a specific view and an adequate interpretation of science, the starting point should be a recognition of the complexity of the dialectic process between theoretical construction and data generation, processing and laboratory analysis procedures; also a recognition of the central role of subjects as interpreters in designing and using scientific representations. I also argue that the *family resemblance* which exists between the constructivist/structuralist empiricism and American pragmatism suggests new avenues for analysing the decision-making process and the role played by subjects who interpret, construct or use models in scientific contexts. A connexion with the pragmatists' thesis and perspective that is very much present, not only in van Fraassen's most recent texts on scientific representation, as some other authors maintain, but also from the outset in his earliest publications.

Keywords Scientific representation • Constructive empiricism • Empiricist structuralism • Interpretation of scientific activity • Pragmatism

I. Perdomo (✉)
Faculty of Philosophy, University of La Laguna, Guajara Campus,
w/n, 38296 La Laguna, Tenerife, Spain
e-mail: inmaperdomo@gmail.com; m.perdomo@ull.es

W.J. Gonzalez (ed.), *Bas van Fraassen's Approach to Representation and Models in Science*, Synthese Library 368, DOI 10.1007/978-94-007-7838-2_2,
© Springer Science+Business Media Dordrecht 2014

Thirty years after the publication of the celebrated text by Bas C. van Fraassen, *The Scientific Image*, philosophical debates still rage regarding his work, his epistemological position and the nature of his proposal: constructivist empiricism, in critical open dialogue with realisms (both old and new) and instrumentalisms. New impetus was added to the debate by the publication of his most recent book, *Scientific Representation*, in which he qualifies some of his basic suppositions and proposes a new name for his empiricism: *empiricist structuralism*. Our aim will be to explore the principal traits of van Fraassen's philosophy of science, arguing that it is one of the most comprehensive and appropriate views of scientific activity, and that the link between renewed empiricism[1] and pragmatism is very close, not only in the last text, as some other authors maintain,[2] but right from his earliest publications.

2.1 What Is Philosophy of Science?

Philosophy of science plays the part of authorised interpreter of the scientific practices, the epistemological orientations that guide scientific procedure, and of the attitudes towards science. Thus, it compounds a vision of science with the aim of understanding human cognoscitive activity at its most articulate, sophisticated and successful. Philosophy is *interpretation*, it proposes an interpretation of science,[3] with the aim of gaining a better understanding of the whole process, activity or body of knowledge, which can only be achieved through dialogue between those involved in the undertaking. Explanatory success or failure is also linked to the agreement reached between the participants in this dialogue, both in relation to the classification of facts and with regard to the assessment of their relevance and meaning. Participants in philosophical dialogue share a common starting point and have, or may establish, a series of basic agreements and values which stem from the culture and historic moment to which they belong. The empiricist-constructivist interpretation of science offers a view of science, a concept of this activity, which is consistent with this fact: science is a greatly admired intellectual undertaking, the paradigm of rational research, but it is also subject to severe criticism in order to avoid dogmatic establishment in any body of knowledge which, by definition,

---

[1] This is how I defined it in the text analysing the work and focus of van Fraassen. Perdomo and Sánchez (2003).

[2] M. Suárez believes that, due to this change of course, van Fraassen ends up "in no man's land. Or in someone else's land. I think we end up in the land of pragmatism." In my opinion, constructivist empiricism and pragmatism have always shared common ground. See Ladyman et al. (2011) (Nov. 2010).

[3] U. Moulines has affirmed that "the philosophy of science constructs interpretative philosophical frameworks which enable us to understand these interpretative frameworks of the reality which we call scientific theory." Beyond the limits imposed by the descriptive/prescriptive dichotomy for defining the task of philosophy of science, what this implies is the possibility of offering a view of things, a way of thinking about certain phenomena in a certain manner. Moulines (1995, 110). This approach is very similar to that offered by van Fraassen.

will always be partial and tentative; thus, we avoid both our tendency to indulge in the realist convenience of belief in an underlying order which our science is just capable of glimpsing, and the dissolution of rules and guidelines in the network of interests and ideologies which plague scientific communities.

In his work, van Fraassen asserts that scientists commit themselves to participating in the search for empirical adequacy. It is an open question as to whether, as individuals, scientists believe that accepted theories are correct, that their work will lead them to discover God's creation plan, that they are on the path to discovering the laws of nature, that their experiments will enable them to discern the structure of certain unobservable entities in whose existence they nevertheless believe. Therefore, the idea that scientists are searching more for empirical adequacy than the truth, or any approach to it, is a question that is compatible with the opinions or beliefs of the individual scientists themselves (van Fraassen 1994c, 181). Scientists participate in a common undertaking, an undertaking in which they establish the empirical adequacy of the theories they produce as the criterion for success, although other criteria may also be defined as relevant. Philosophy of science explains this by analysing the objectives of science, as reflected in the practices and values designed and sustained by the scientific community itself, the beliefs and opinions implied in the acceptance of certain theories, the intentional aspects and the use of scientific models to represent and explain phenomena and the processes of measurement, simulation and technological development, which form the basis of the theoretical construction process.

Style defines the special character or means of expressing concepts that an author bestows on his or her work (van Fraassen & Sigman 1993). Applied mainly to artistic activities,[4] this concept is equally valid for illustrating the character which van Fraassen lends to philosophy of science. The concept of style immediately suggests that of creative imagination and interpretation, and in the case of philosophy, this also translates into conceptual elaboration, the ability to imagine and create new categories or concepts which enable us to illustrate or interpret the specific characteristics of the object in question, in this case scientific activity, and the processes and results of said activity. A philosophical style defines the questions which make up its central focus, as well as the rules or criteria with which the results are assessed or appraised, success and productivity criteria and aesthetics, etc. It also reveals attitudes to topics associated with this activity: a theory regarding how facts are constructed in the laboratory, how data and theory mesh, how theoretical models are used to respond to questions defined as relevant in a specific historical context, referring to certain questions which are pertinent to the philosophy of science but which, above all, offer a vision, an approach, a specific "lens" which enables us to shed light on certain shady areas from other alternative approaches.

---

[4] It is also applied fruitfully to the analysis of the history of science, as a means of putting into practice different styles of scientific reasoning and creative imagination. The history of science is understood as the result of applying different styles of scientific thinking, and as the product of both processes of mutation and the continuity of said styles of thought. This is the approach adopted by A. C. Crombie (1994).

It may perhaps seem that the inevitable result to which this line of thought leads is an admission of the existence of a multitude of approaches, all at the same level, each one with its own specific set of values, criteria and preferred topics, all equally consistent: indeed, this is what post-modern epistemological thinking would have us admit. And it is true that, having reached this point, it is indeed the only way out; the only possible coherence is the internal coherence of each approach or perspective. However, van Fraassen argues that, before reaching this point, the study of philosophy of science and the discipline itself should adopt as its starting point a sceptical, self-critical and empiricist attitude. Strictly speaking, there is a plethora of approaches, but only two attitudes on which to base analysis and conceptual development: one based on received wisdom, and the other one sceptical, empiricist and critical.[5] It is this second one which enables us to carry out our interpretative task unencumbered and ensure a philosophy of science committed to the task of interpreting the complexity, sophistication and contextual nature of the construction, assessment and use of scientific knowledge.

B. C. van Fraassen assigns philosophy an important role as the interpreter of the interpretations of the world,[6] and this implies a complete renewal of empiricism; from his initial work in 1980, *The Scientific Image*, to his most recent offering in 2008, *Scientific Representation*, he confers on philosophy of science a distinct, key role which is a far cry from its traditional normative and justificatory approach. B. C. van Fraassen's constructivist empiricism defines scientific practice as that which enables the proliferation of interpretations, the suggestion of different models ordering, measuring and interpreting both phenomena and the philosophical task itself, as an interpretation of this interpretative action. This empiricist approach is, in my opinion, also similar to that adopted by H. Longino, who argues that the values which guide the different interpretations are contextual and historical cognitive values, both in science and philosophy, and defines this view of scientific knowledge as contextual empiricism, in the following terms: "It is empiricist in treating experience as the basis of knowledge claims in the sciences. It is contextual in its insistence on the relevance of context — both the context of assumptions that supports reasoning and the social and cultural context that supports scientific inquiry — to the construction of knowledge." (Longino 1990, 219.)

An adequate analysis, both in the world of the basic experience of science and in that of the investigating subject and communities of scientists, and the handling of

---

[5] There are other models we could use to illustrate this attitude: the attitude of the feminist critique of science, for example, and more specifically, that of critical and contextual empiricism, defended by H. Longino, for whom the possibility of a future non-androcentric science is necessarily based not on the absolute condemnation of science, but rather on the adoption of a critical attitude to both contextual values and internal methodological criteria and the rules that define this practice (H. Longino 2002). This attitude is also expressed by Kant in *Prolegomena*, when he confesses that Hume interrupted his dogmatic slumbering, giving his research a completely different character. This is, according to van Fraassen, a perfect illustration of the empiricist attitude, although Kant did not define it as such.

[6] van Fraassen explores this idea of interpretation, which is similar to that used in the arts context, in "Interpretation in science and in the arts," 1993, 73–99.

adequate notions applied to the description of the processes involved in the construction of knowledge, imply the defence of this empiricism as a global approach, to the extent that it illustrates the type of interactive, interpretative and constructive process which takes place between the epistemic community and reality.

2.2   The Semantic Conception of Theories and Constructivist Empiricism. Scientific Activity as a Constructive and Intervening Process

van Fraassen's empiricist approach was developed within the semantic conception of theories, which offers a basic approach for philosophy of science's new agenda following the foundationalist failures. It conceived scientific theories as sets of models, and opted to formalise them following semantic methods. However, it also analysed the relationships existing between theories and the epistemic community (i.e., subjects, active agents in the process of exploring and intervening in the world), the processes of accepting or rejecting theories and the active role of experimentation in the construction and development of theories; although it is also true that it attached less importance than other similar approaches[7] to the role of prior theories and the processes of scientific change. van Fraassen's approach, within the framework of the semantic conception, enables us to navigate around that which, in my opinion, constitutes the core of debates about science: scientific activity as a constructive and intervening process which generates interpretations of the world. The debate regarding the role of the decisions made by scientists, their commitments to theoretical frameworks which are considered "expert guides" in the development of the scientific image of the world, as well as the foundations of theoretical acceptance and epistemological stances and attitudes to science. Questions which demand that which van Fraassen calls the *self-location* of subjects in relation to the body of knowledge, similar to the process of the *self-location* of the user in relation to a map which tells them where they are, an issue we will deal with later on.

---

[7] I am referring to Balzer, Sneed and Stegmuller's structuralist view. The structuralist approach defended by both perspectives provides a set of conceptual tools for dealing with the fact that science is, above all, a kind of activity whose aim is to provide an interpretation of its object of study in terms of its structure. They defend this activity as being essentially constructive in nature, i.e., scientists construct models, mathematical objects, which are then used to represent nature. Structuralism continues to defend the ideal of axiomatisation, opting for mathematical methods such as set theory to develop its vision of science. Thus, it offers a series of tools appropriate for reconstructing highly mathematised theories, enabling the adequate establishment of the set of elements and relationships which make up a theory, as well as the relationships between different theoretical elements, whether they be contemporary or part of a historic series. However, at the same time, in our opinion, this approach was unable to offer an image of the processes of theoretical construction based on the idealisation of the world of experience, and therefore, an adequate image of the relationships existing between theories and the world, issues which van Fraassen's approach does tackle.

Scientific activity is not simply a process of discovering truths, no matter how approximate or fallible these truths may be; rather, it consists of a process of constructing appropriate models for explaining phenomena which have conveniently been idealised by the procedures which make up "laboratory life" Theories are conceived as sets of models, extra-linguistic entities which enable scientists to represent, to explain and to intervene in the world; in short, they enable their use, in general terms, for a wide range of different purposes. Scientific theories focus on a type of phenomenon which constitutes their *intended scope*[8] and the aim of every theory is to present a general description of these phenomena which can be used to satisfy the demand for explanation, prediction and detailed description. To this end, the theory abstracts certain parameters from these phenomena, minimising their excessive complexity. These parameters are those deemed by scientists to be relevant, and the supposition is that it is they alone which have any influence, and that therefore, phenomena are isolated systems[9] that can be defined and described solely on the basis of those parameters selected by the theory. Thus, the theory characterises not the phenomena which fall within its scope, but rather ideal copies of said phenomena: *physical systems*.

A physical system is not a system of real phenomena, but rather a highly idealised copy of real phenomena. Thus, although the field of application of a theory is a phenomenon domain, or a specific type of phenomenon, and we can offer explanations based on that theory, the determination of these phenomena is carried out on the basis of a series of parameters abstracted from them, which have been idealised and selected by the theory itself, or to be more precise, by scientists, in accordance with the aims of the research, with only some of the many parameters involved in complex real phenomena being chosen. Thus, the theory constructs an idealised, counter-factual copy of the phenomenon system, which assumes that only those aspects selected actually intervene. This is a constructive element which enables scientists to establish how phenomena would behave under these ideal conditions. The universe of science, in this sense, is not the complex world of events, but rather that of experimental and laboratory research in which said selection takes place.

From his constructivist empiricist approach, van Fraassen believes that this idealisation is not carried out directly by the theory itself, but rather by a theory of experiment which, based on experimental data and measurement reports, etc., constructs data models called *appearances* (van Fraassen 1976, 631), which may be considered descriptions of phenomena relevant for the theory. In this case, as we shall see later on, the idealisation is increased, or even doubled, by this step through a theory of experiment. Physical systems or appearances are also considered to be isolated, and

---

[8] This is a concept used by F. Suppe (1974/1977), 257.

[9] This *fiction of isolation* is the reason why the results obtained are, strictly speaking, false. It is, on the other hand, the reason for the explanatory and predictive force of the hypotheses, hypotheses which rather than talking about how phenomena behave, focus instead on how they would behave in the event of said ideal conditions coming to pass. An updated debate based on contemporary references to the Kantian Vaihinger and the philosophy of "as if," or the analyses which explore the use of fictions and simulation in the construction of models and theories.

this itself constitutes another idealisation factor. This *fiction of isolation* is the reason for the theory's lack of precision and the patent falseness of laws when compared to phenomena. The essential function of a law is to describe the behaviour of the type of physical systems which are the focus of a scientific theory; in more specific terms, its function is to describe the conditions of what is physically possible. The difference[10] between *laws of coexistence, laws of succession and laws of interaction* enables scientists to describe the possible states of a system, its trajectories and its behaviour during interaction. Once the laws of the theory have been included, the *state space* is established and the behaviour of a physical system, or the idealisation of a phenomenon, is represented by diverse configurations imposed on the state space in accordance with the laws of the theory, and only those points of the state space whose coordinates satisfy a specific equation will be physically possible.

From this perspective, we could assert that theories are structures, and these structures are state spaces which have a series of specific configurations imposed on them by the theory's laws. In *Laws and Symmetry*, published in 1989, van Fraassen develops this thesis further and asserts that laws are nothing more than the basic principles of a theory, its fundamental equations, model laws. They are those key characteristics by means of which models can be described and classified; and it cannot be claimed that these laws correspond to the laws of nature, as the vast majority of philosophical tradition has established (van Fraassen 1989a). *Theoretical definition* specifies a family of structures which are theoretical models. *Theoretical hypothesis* reflects the affirmations of the theory regarding the real world, i.e., the affirmations that certain real, or at least observable, systems belong to the defined class, since these abstract objects constructed by theoretical definition are related to appropriately mathematised and idealised physical objects. While in that related to theoretical definition there is almost unanimous agreement between all followers of the semantic conception, in that related to theoretical hypothesis and the specific relationship between theory and the world to which it applies, opinions are divided. A number of different stances have been adopted, although the two most commonly debated alternatives are: constructive realism and constructivist empiricism, whose vision is as follows:

> To present a theory is to specify a family of structures, its models; and secondly, to specify certain parts of those models (the empirical substructures) as candidates for the direct representation of observable phenomena. The structures which can be described in experimental and measurement reports we can call appearances: the theory is empirically adequate if it has some model such that all appearances are isomorphic to empirical substructures of that model. (van Fraassen 1980, 64.)

Theories only aim to be empirically adequate. However, the empirical adequacy of a theory is only affirmed after a process of deliberate selection which begins with the routine task of processing the enormous amounts of data generated by measurement and observation instruments. The demand for adequacy is firstly, a structural demand, i.e., it is a relationship between a data model and a theoretical model. It is a

---

[10] The difference is defined by van Fraassen in "On the extension of Beth's semantics of physical theories," 1970, 325–339.

mathematical relationship. However, it is also an affirmation of adequacy in relation to the structure of the real phenomena described in terms of the theory's relevant parameters. This means that observable phenomena, even if they are only instrument readings, are observable by anyone, but the way in which they are described by scientists (human beings who defend previously accepted theories and who make assumptions and have values and options) may differ widely. The empirical infra-determination of all theories, but particularly the fact that all descriptions of nature are theoretically heavily conditioned, means that the defence of a view of science as an interpretative activity makes perfect sense.

Since the publication of *The Scientific Image* and other previous works, Bas C. van Fraassen has defended his constructivist empiricism as the most appropriate philosophical interpretation of scientific activity. This empiricism is gradually defined also as it dialogically confronts scientific realisms and the new minimal realisms which admit the fallibilism, approximation and tentative postulating of "behind the scenes" observational entities or processes, but which do not renounce to the "metaphysical instinct" of the postulation of entities as real causes of the processes being explained. Reality based on explanatory and predictive success. Concepts such that of "laws of nature," or the natural principles captured by our best theories can be renounced, but not the idea of need which gives meaning to our notions of causality and explanation. This, at least, is what R. Giere argues (Giere 1999). In particular, the core of what van Fraassen defines as metaphysical ingredients of realist philosophical positions consists of giving absolute priority to the demands of explanation and satisfying them through explanations via postulation: In other words, explanations which postulate the reality of certain entities or aspects of the world, which are not empirically evident. For van Fraassen, "science aims to give us theories which are empirically adequate; and acceptance of a theory involves as belief only that it is empirically adequate." [11]

Indeed, in all his works, van Fraassen claims that any other virtue required of a theory, above and beyond its empirical adequacy, is always pragmatic. This does not make the theory more adequate or approximately true, only preferable. These preferences, it could be claimed, may be based on interests, tastes, better efficiency, adequacy to research objectives or technological performance. All this forms part of the series of reasons for which we opt for one theory or another, says van Fraassen; acceptance has a pragmatic dimension. And,

> To accept a theory is to make a commitment, a commitment to the further confrontation of new phenomena within the framework of that theory, a commitment to a research programme, and a wager that all relevant phenomena can be accounted for without giving up that theory. (…) Commitments are not true or false; they are vindicated or not vindicated in the course of human history. (van Fraassen 1980, 88.)

---

[11] van Fraassen (1980, 12). Although in other later texts van Fraassen tackles the question of belief not as an all or nothing issue, but rather by incorporating the probabilistic model. Belief, according to W. James, as van Fraassen read him, is a question of will and is, above all, a decision to make a commitment.

This empiricism is therefore defended also as an attitude, that which outlines a certain approach to factual questions as being paradigmatically rational. This concept of rationality is written in lower case. In other words, it is a "permissive" concept of rationality. It is a minimal and instrumental rationality which only advises us not to sabotage our chances of defending and justifying our commitment to a specific interpretative framework. However, this commitment includes an element of free choice or voluntarism,[12] which cannot be understood as the mere modification of a previously-held opinion "in the face of new evidence," since this concept has also been clearly reinterpreted in light of current scientific practice. Scientists commit to a specific theoretical framework providing they believe that this is the best way to achieve the objectives established within the community to which they belong. The choice of one specific option from among other possible ones, in order to offer an adequate interpretation of phenomena, leads us in a certain direction; the choice implies commitment, the implicit selection of certain parameters as relevant and the involvement of certain values and assumptions. But the initial position of empirical risk is maintained right up to the end (van Fraassen 1989a, 261), since phenomena can also be modelled on the basis of alternative symmetry arguments.

This prompts van Fraassen to call into serious question the efforts made to formulate an adequate idea of scientific law associated with that of need and universality, a reflection of the principles of order which truly exist or the laws of nature.[13] Particularly, any image of science presented in this way as a mere representative activity overlooks, as Hacking (mainly in 1983/1996) also reminds us, the fact that it is, at heart, an intervening practice. In fact, the dialectic relationship between theory and experiment, as we see it, constitutes the core of theoretical construction, but also of technological innovation.

It is obvious that if philosophy aims to offer a specific view and an adequate interpretation of science, the starting point should be a recognition of the complexity of this dialectic process between theoretical construction and data generation, processing and laboratory analysis procedures. Received topics, arguments which illustrate our faith in a world order which our theories reflect, the emphasis on the explanatory task of science, the central nature of notions of law, causality and evidence are the old dreams of a philosophy of science which is well past its sell-by date, and are revealed as totally anachronistic when we turn our gaze to examine the heart of scientific activity: laboratories or *large scientific facilities* filled with observational and experimental equipment.

The construction of "appearances" to use van Fraassen's term, or "physical systems" as F. Suppe's calls them, or simply, and in general terms, phenomena which have been idealised enough to be treated scientifically, is increasingly restricted to the laboratory field or to large scientific facilities, since even a discipline such as astronomy has stopped being strictly observational and has become a discipline which processes, simulates or *deforms* light so as to obtain images which interpret

---

[12] The notion is recovered by van Fraassen from American pragmatism, particularly from the works of W. James. It is evident in his text from 1897/2003. Vid. also Perdomo (2003).

[13] The arguments are mainly developed in van Fraassen (1989a), passim.

what is observed in terms of the theoretical framework to which the scientist in question is committed. This relationship is not unidirectional, but rather dialectic, since constant feedback is produced between the experimental and theoretical levels. We can therefore talk about mutual conformation aimed at satisfying pre-established objectives. Data, instruments and ideas are gradually adjusted in a kind of symbiosis resulting from the deliberate process of selection, demand for, and invention of new instruments designed to generate data which will enable the development of a theoretical hypothesis, while at the same time opening up new areas of experimental development (Hacking 1991, 29–64). These studies show the complex interactions between these different elements, between ideas (be they theoretical, systematic or hypotheses) or theories regarding the working of apparatus or things, i.e., all technical instruments, sample preparations, detectors, data generators, etc. and the world of generated, assessed, analysed and, finally, interpreted data. The gradual symbiosis between theories and laboratory equipment is a fact in mature science; they evolve towards mutual adjustment, to the point at which it is possible to stop generating data which are not relevant to theoretical hypotheses. Measurement, van Fraassen also affirms, is designed to answer specific questions, and the information derived from the measurement outcomes is relevant to the responses provided.

However, this symbiosis and internal coherence, which generate a certain degree of stability which is nevertheless contingent, imply that the variation of one element may destroy everything else. Or, to put it another way, alternative data[14] may be produced, data which are generated due to the stagnation and review of practices, to alternative research teams with different values or to the application of more powerful instruments which generate new kinds of data which cannot be accommodated within the previous theoretical framework. The important point here is that, in this case, the incommensurability of both the old theory and the new one which interprets these new data is radical, since we are no longer talking about theoretical or semantic incommensurability, but rather incommensurability which is produced at the level of the instruments used and the data generated, which cannot be interpreted or accommodated by the previous framework. Despite this, however, the old theory may continue to work perfectly in its own data domain, which provokes a curious image of the diversity and locality of science.[15] This diversity is mainly the result of the laboratory production of phenomena using different techniques and instruments.

As defining characteristics of science, constructivism, symbiosis, contingence and diversity provide a new image of scientific activity in which experience,

---

[14]These data may arise in what have been dubbed the "margins of science" The similarly to Feyerabend is evident, but the resemblance to new studies of science from the gender perspective is also patent. These studies have levelled radical criticism at many aspects and ideas of the more traditional philosophy of science and the resulting images of science, while at the same time outlining new epistemological proposals.

[15]The resulting image may be that of a patchwork of theories, disciplines and laws, with no hierarchical order or systematic relationship. Vid. N. Cartwright (1999).

interpretation and transforming action become key concepts. It is for this reason that the connection between empiricism, constructivism and pragmatism is the one which, in my opinion, offers the best interpretation of that activity.

2.3 Constructivist Empiricism and Pragmatism

A careful analysis of the "family resemblance" which exists between the constructivist empiricism defended by Bas C. van Fraassen and American pragmatism suggests new avenues for analysing the decision-making process and the role played by the subjects who interpret, construct or use models in scientific contexts. Concepts recovered from the pragmatism of W. James, such as voluntarism and the idea of the conflict between epistemic human desires to believe in the truth and to avoid errors are used by van Fraassen to mitigate the rigid proposals of the Bayesian or evidential theories of decision. van Fraassen chooses to view the acceptance of theories as an open, tentative process, in which epistemic agents decide to adopt a theory as their "expert guide," in order to continue moving towards the construction of the model-theory. In his work, van Fraassen has developed other concepts and approaches with pragmatist leanings, such as his pragmatic theory of explanation, or his concept of the ongoing dialectic between theoretical development and experimentation as the key to the process of theoretical construction. These are only some of the aspects which align him with the thesis of pragmatism, or, to put it in a slightly different way, the renovation of empiricism carried out by pragmatism is perfectly illustrated in van Fraassen's work.

Let us not forget that the initial convergence between the pragmatic trend, particularly as developed by Dewey, and logical empiricism at the beginning of the twentieth century was diluted by the academisation of the logical-empiricist trend and the abandonment of committed social discourse by empiricists from the *Aufbau*[16] culture, just as C. Morris recommended to the old members of the Circle, now installed in American universities following their exile. Both the philosophy of logical empiricism in the context of the *Aufbau* and the philosophy of Dewey were motivated by the technological triumph of science and claimed for science also the capacity of transformation. Neurath's rejection of metaphysics also implied a political conviction of the advent of a liberating, modernist and rationalist social movement. The social benefits of *scientific philosophy* were a common cause for

---

[16] The political, cultural and social context of the inter-war period, in which the Vienna Circle and the Berlin Group arose, has been widely studied by intellectual and political historians. In his work, P. Galison presents what he terms the *Aufbau* culture. The concept has been badly translated as "reconstruction," an interpretation which dilutes all its original revolutionary meaning. The original authors used the term to express a radical sense of newness, a breaking away from the past and a deep-rooted conviction that the inauguration of a "new world" should not be superficial, but should rather mean a complete transformation of culture, education and architecture, expressed in the Bauhaus movement and the new ways of reasoning. Galison (1996).

concern among empiricists and pragmatists like Dewey, for whom the reworking of classical empiricism meant the replacement of past experience with future experience, as the basis of the cognoscitive process.

The formal encapsulation of logical empiricism, a stimulating philosophical project which had much in common with pragmatism,[17] resulted in a specialist academic discipline of philosophy of science, which Putman baptised during the 1960s as the *received view*. In Galison's opinion, the *Aufbau* culture did not cross the Atlantic, and during the 1950s, the majority of philosophers in the American context believed that pragmatism was "wrong" and logical empiricism "right," and often cited the crossfire of declarations between Russell and Dewey: whereas Russell believed that Dewey's pragmatism was nothing more than American commercialism disguised in philosophical garb, Dewey was convinced that Russell's dry, technical philosophy was nothing more than the expression of decadent, aristocratic, English sensibility.

I. Hacking (1983/1996, 62–69) defined van Fraassen as the new defender of positivism, following in the footsteps of Hume during the mid eighteenth century, Comte during the 1830s and the advocates of logical empiricism from the 1920s to the 1940s. Hacking underscored the series of theses which define this position and which are, in his opinion, common to all the aforementioned authors: the verificationist ideal, the negation of causality beyond the mere verification of regularity, or the rejection of the idea of entities whose existence is adduced indirectly, through the postulation of dubious causes or explanations; together, all this constitutes the positivist commitment to "opposition to metaphysics" Despite locating van Fraassen in this trend, his style is characteristic of precisely all that which denies dogmatic establishment in any stance and which defends a constant critical, sceptical attitude — the hallmark of constructivist empiricism. This empiricism is one which maintains some of the assumptions which characterise this trend, not from the eighteenth century onwards, but from as far back as the nominalism of the fourteenth century, as van Fraassen himself points out (van Fraassen 2002, 1994b), but which is considerably far removed from the academic logical empiricism developed in the American universities from 1930 to 1960.

van Fraassen's constructivist empiricism also owes something to pragmatic postulates. Pragmatism, whether it be Peirce's version or in the path followed by James, Dewey, Lewis or Rorty, is antirealist. The concept of truth is radically redefined. It can be conceived as either the end product of the efforts of a community of researchers pursuing a specific goal, or as a set of acceptable general conclusions. Emphasis is placed on the method and on the end result of its application, as Peirce argues, or on

---

[17] Richardson's analysis moves away from specific philosophical theses in order to focus on the philosophical commitments, goals and aspirations of empiricists and pragmatists, on the motivational and attitudinal elements of *scientific philosophy*, a project shared by both parties in an attempt to overcome an aging philosophy closely allied to traditional conservative discourses. From this perspective, the convergence between empiricism and pragmatism becomes much clearer. Richardson (2002).

the process of constituting knowledge on the basis of our experiences, as James and Dewey claim, thus turning truth into guaranteed acceptability.

Thus, just as James rejected absolute scepticism, asserting that we are capable of establishing truths about ourselves and about what the world is like, so van Fraassen also affirms that in relation to what is observable, in relation to what we have empirical access to, it is possible to assert the truth; however, equally, and contrary to the other extreme represented by absolutism or dogmatisms, both authors argue the fallibilism inherent in all demand for knowledge. We cannot attain objective certainty or absolute guaranty. It is in the rejection of both stances that the virtue of the empiricist perspective lies: experience is the only legitimate source of our opinions about facts. And therefore, all conclusions regarding issues of fact are susceptible to modification in light of future experience. "In this way, theories become instruments rather than answers to enigmas upon which we can rely. We must not lie back and relax on them, but rather move forwards and, on occasions, with their help, rethink their very nature." (James 1907/1997, 41.)

In pragmatic terms, knowing is equivalent to bringing a series of skills to bear on an action aimed at a specific purpose, without forgetting that both are dynamic and moreover, will be subject to different kinds of feedback as a result of the research itself. This implies a radical rethinking of reality itself, of our access to it and of the concept of experience and knowledge, an approach which would be impossible without another basic category: interpretation. Reality is no longer a non-problematic *factum* and accessibility to it inevitably implies a subject with purposes and the capacity to act, whose context is a scenario, a world of experiences, from which said reality is critically elucidated. This critical elucidation of reality therefore implies the acknowledgement of the active role of the subject in the conformation of a cognoscible reality.[18]

The role of the subject is vital to the process of theoretical construction; observation and reasoning are not objective, neutral activities, but are rather mediated by the contexts and criteria of scientificity established by the scientific community itself, interpretation occurs at different levels, the responses provided to demands for explanation are contextually relevant and research objectives are designed in close alignment with applicative objectives. In short, models are used by subjects to attain planned objectives. And all this presupposes a view of scientific activities which further strengthens the connection between constructivist empiricism and pragmatism. The masterly analysis of scientific representation offered by van Fraassen in his text *Scientific Representation* perfectly illustrates the connection with pragmatism, a connection which is even closer here than in his previous works and which links empiricism with the use of models to represent the world of experience, in order to target our actions towards the goals to be achieved.

> A view of science would hardly be empiricist if it ignored the uses of science, as a resource for praxis. How are theories and models drawn on to communicate information about what thing are like, to guide our expectations in practical affairs, to design instruments and technological devices, to find our way around in the world? (van Fraassen 2008, 88.)

---

[18] These ideas are developed in more detail by Ángel M. Faerna (1996).

2.4 The Scientific Representation of Reality. Constructivism, Interpretation and Uses

Many philosophical texts on scientific representation have been written over recent years. The same question crops up time and time again: despite the levels of idealisation, constructivism and interpretation inherent in scientific practice, how do theories connect to the world? Models should reflect real, significant aspects of the phenomena being studied, even if only in terms of their structure; this has also meant a new revitalisation of structuralism (see Psillos 2006. Also Brading and Landry 2006). Classical or traditional analyses of representation focus on the similarities between aspects of the model and aspects of reality. Precision and completeness are usually presented as the principal values associated with the act of representation, but we must first admit that this is a question of degree. And, in relation to either value, is also required a context in which decisions can be made regarding which aspects to select and which criteria to apply. Thus, representation should be defined as an intentional activity, subject to assessment and the application of criteria, and relative to the context of use and production. However, it is also common to start by establishing a description of representation in the field of the arts, and to analogically transfer the conclusions reached to the field of science.

Thus, questions of similarity or resemblance are posed at the argumentative core of the issue of representation, although the analysis may be rendered even more complex if notions of perspective, distortion or even fiction[19] are introduced into the heart of the debate. In this sense, the profusion of details and examples provided by van Fraassen in his texts on scientific representation are immensely enlightening. van Fraassen coincides with M. Suárez in affirming (Suárez 2004, 771) that representation is not the type of notion that requires a theory to elucidate it, that there are no necessary and sufficient conditions for it, and that the most we can do is describe its more general characteristics. What is a representation? How exactly does it represent? What are the essential elements for talking about an adequate representation? And what are the conditions of possibility for scientific representation, or its variants. These are questions which van Fraassen tackles with skill and dexterity in his text. The responses centre around one key issue: the crucial role played by use and practice, in a new approach to the core of pragmatist thinking. "There is no representation except in the sense that some things are used, made, or taken to represent some things as thus or so." (van Fraassen 2008, 23.) The *Hauptsatz,* term used by van Fraassen, of the text could have been written by pragmatist philosophers, for whom being in possession of a theory or representation of reality means being in possession of a practice, of a connection between actions and ends, symbolically mediated by a system of representation which bestows sense and meaning and which functions in this area of experience.

---

[19] A comprehensive study of the role of fictions in the construction of models and theories and the epistemological consequences of the use of these strategies has been edited by Suárez (2009).

In *Scientific Representation*, van Fraassen presents a multitude of examples demonstrating that, in many cases, it is not the model of the reflection, but rather that of the diffraction, so to speak, that constitutes the basis of successful representation. As in caricatures, which highlight a face's most characteristic features, distortion also plays a role in representations. Sculptors distort harmonious proportions in order to ensure that they maintain certain forms from a certain distance and angle, and painters calculate perspective in order to draw figures of the size appropriate for representing the relative distances between elements. In the field of advanced science, the adaptive distortable optics of the new great telescopes, such as the GTC, enable light to be distorted in order to "eliminate" the aberrations caused by atmospheric perturbations. The front of the wave is analysed first by a sensor which determines its aberrations. This information is sent to the phase reconstructor, which calculates the corrections to be made and the distortions the distortable mirror must adopt in order to compensate for the original aberrations detected at the front of the wave. The result is a much clearer image which is, according to researchers, more or less equivalent to what we would see from space. Although in fact, what astronomers are actually doing thanks to this technology is generating images of *how the object should appear* if the theory which interprets it is correct.

In the example of the painter, the representation achieved by mathematically calculating the correct perspective is adequate only in relation to the values appreciated from the Renaissance onwards. Paintings from before the *Quattrocento* reflect the size of the figures in relation to their importance in the scene, rather than relative to the logic of spatial relations and perspective. In fact, when we observe these representations, we need to be aware of these codes and values of representation in order to interpret the paintings correctly. Thus, a representation is an adequate representation of whatever only in relation to a representational system which covers such a case and which confers upon it its ultimate meaning. Similarly, the images of the universe constructed by large telescopes enable representations of the universe which can only be interpreted using the techniques and theoretical models used for that purpose.

According to van Fraassen, we really should distinguish between *representation of* and *representation as*, and the latter cannot be conceptually reduced to the former, since although the former is not without interpretative elements, interpretation is central to the latter (van Fraassen 1994a). The simplicity of the idea of mere geometrical projection, argues van Fraassen, is lost. *Representation as* is constructed and this construction is not unique; the same aspect can be represented in various ways, since the behaviour of the phenomena in question allows for different interpretations. Something is represented as this or that, and during this process we gain an understanding of a certain aspect of the phenomena; in other words, appropriate comparisons have the virtue of facilitating understanding.

> There is no such thing as 'representation in nature' or 'representation tout court;' the question whether one given object is a representation of another is an incomplete question. Specifically, in science, models are used to represent nature, used by us, and of the many possible ways to use them, the actual way matters and fixes the relevant relation between model and nature. Relevant, that is, to the evaluation as well as application of that theory. (van Fraassen 1997, 523.)

Relevant relationships between models and the world: This is a vital aspect of scientific practice, and enables us to approach its analysis from a more pragmatic perspective. Both the selection of the aspects of the model chosen to represent reality thanks to their definition as similar, and the decision as to whether or not the similarity expressed is sufficient, may depend on the purposes for which the model is being designed and applied. In other words, it is a function of the context of use, rather than of the mere relationship between the model and reality. Representation fulfils its function only if we accept a certain interpretation based on a series of codes of acknowledgement (visual, symbolic, cultural, etc.), which we accept as valid or adequate, with which we share a way of seeing and perceiving the world and which enable us to act. The level of constructivism of these codes is very high. However, moreover, representation also implies the *intentionality* of the agents as a vital element.

Nelson Goodman (1976, 33) tells a story in which, in response to a complaint by the playwright Gertrude Stein that her, now famous, portrait looks nothing like her, Picasso responds by saying "no matter, it will." It is obvious that, being aware of his artistic authority, Picasso knew that it would end up determining the "represented object" in the conventional manner. If the painter claimed that the figure was Stein, then all "informed" subjects would accept that it was so. The story of the portrait is actually even more interesting, since the different ideas regarding Stein's representation suggest other possible interpretations, such as, for example, that in fact, rather than a portrait of Stein's actual physical features, what Picasso painted was a portrait of her personality traits. In other words, Stein's strong character and vanity was represented by Picasso in the form of a series of physical features and a specific expression, which observers may perceive as an adequate representation of the playwright, since they recognise the physical features conventionally associated with these psychological traits within a shared set of codes. Another possible interpretation is that the figure of Stein actually represents the couple; it is a kind of merging of the features of Gertrude and Alice, recognisable to those who were aware of the relationship. We could even propose a new interpretation, i.e., that just as Stein developed a narrative style far removed from convention, inspired by the teachings of W. James himself, in which the plot was almost entirely eliminated and the prose was free and radically innovative as regards syntax and punctuation, so Picasso did the same in his pictorial representation of the playwright. Basically, he was experimenting with the possibilities of the artistic language, establishing new interpretative codes for reality.

Nevertheless, no matter how interesting this line of argument may be, we should stop here and remember that, despite all the comparative analyses and suggestive analogies that can be established between representation in art and literature and scientific representation, the latter has its own specific traits.[20] Scientific theories, presented through their set of models, are abstract, mathematical structures, and in this sense, the structuralist concept associated with the new label "structuralist empiricism" refers to the theory that all scientific representation is basically mathematical in

---

[20]This was argued also by Steven French (2003).

nature, and according to van Fraassen, this is a theory not about what reality is like, but rather what science is like.[21] Therefore, the question remains the same: how can an abstract entity, such as a mathematical structure, represent something which is not abstract, like something from nature?

van Fraassen invites us to break down the question by examining the process by which scientific representations are constructed; a perspective which sheds light on their internal elements and dynamics. It is a perspective which is radically different from the usual analyses of representation, which focus on analysing representations as finished products, examining their adequacy or looking for the keys of the representational relationship between theoretical models and the world. From this synchronous analytical perspective, the classification and description of alternative analyses of representation tackle only one aspect of the issue. M. Suárez asserts that van Fraassen defends an intentional concept of representation in which the relationship to be established between representation and that which is represented is one of isomorphism. According to this author, the demand for isomorphism is established between the empirical substructures and the observable part of the world, which implies the defence, in his opinion, of "the view that scientific representation is isomorphism."[22] However, it is important to differentiate between observable phenomena and appearances, and this clarification implies, in his opinion, the introduction of a triadic model: theory-phenomena-appearances, motivated also by van Fraassen's closer attention to the practices of measurement and instrumentation, characteristic of contemporary science, and to the questions of how models are used. These new ideas are, claims the author, presented by van Fraassen in his latest text, and imply the justification of the transformation of constructivist empiricism into structural empiricism. Suárez concludes that, as a result: "The theory is then empirically adequate if it embeds the appearances — and this no longer carries the implication that a substructure of the theory must be shown to be isomorphic to the phenomena." (*Ibid.*)

In my opinion, the differentiation between observable phenomena and appearances is one of the most characteristic traits of van Fraassen's proposal, not just in this text, but right from his early work during the 1970s, which was the result of his research into the Copenhagen interpretation of QM. van Fraassen clearly differentiates, as stated above, between *phenomena*: observable entities (objects, events, processes) which can be measured, including the outputs of measurement instruments, and *appearances*: the contents of the observation or the measurement outcomes (determined, therefore, by the type of measurement process or procedure employed and the instruments, etc. used or developed). Phenomena are observable but their "appearances," i.e., how *they appear to us* as the result of a certain type of measurement or observation process, are something different: "the measurement outcome shows not how the phenomena are but how they look." (van Fraassen 2008, 290.) Appearances are structured according to *data models*: "the selective relevant depiction of the

---

[21] van Fraassen (2008, 239).

[22] Ladyman et al. (2011), Scientific representation: A long journey from pragmatics to pragmatics. *Metascience*. Book Symposium, published online, November 2010.

phenomena by the user of the theory required for the possibility of representation of the phenomenon." (*Ibid.*, 253.) Given that they are means of presenting phenomena, appearances are changeable.

The isomorphism relationship is demanded (as ideal) between appearances and the empirical substructures of models, which may offer an adequate theoretical explanation for them in accordance with the established goals, specific problems to be resolved or questions asked. Data models should be able to be *ideally isomorphically embedded* into theoretical models. However, this relationship which is established between two mathematical structures, between data models or appearances and the empirical substructures of the model, does not yet constitute a representation, although it is a prerequisite if we are talking about scientific representation. What is required also is a subject (an individual or group in a context which confers adequate signs and meanings) who expresses the intentionality of said representation. And it is for this reason that a certain Wittgensteinian movement or a recovery of the Kantian lessons occurs, common also to classical pragmatists and empiricists, in which the subject of knowledge becomes an agent who must organise and interpret the experience before extracting knowledge from it. Moreover, the world is not cognoscible without this interpreting subject. Thus, it is clear that the relationship is not dyadic (model-world), but rather triadic and involves the user, and it does so at different levels or moments of the process, not only during the selection of the relevant aspects during the construction of appearance and data models, but rather in an ongoing manner throughout the whole research and model-theory construction process.

As I interpret him, van Fraassen has not changed his position at all regarding that expressed in his earlier texts; he has merely underscored even more the phrase *by the user*, which, I sustain, is a more explicit option in this text than in others due to the theory of pragmatism, but whose content and orientation had already been presented to the constructivist empiricists. By highlighting the role of the user, have we, van Fraassen asks, succumbed to the post-modern belief that nothing exists beyond the text? The answer is obviously no, but the means of tackling the problem implies a Wittgensteinian movement, as he himself affirms (van Fraassen 2008, 254). The relationship between theory and phenomena is a relationship between mathematical structures, between data models and theoretical models, but the structural relationship between the model in question and the phenomenon, described and mathematised in a relevant way for users, is not enough to turn the model into a representation of the phenomenon.

The importance of the interpreting subject in a process of these characteristics is significant, and implies a continuous decision-making process in which values, purposes and criteria play a key role. The process of theoretical construction is highly sophisticated and contains different levels of idealisation, abstraction and constructivism. Constructivist empiricism explains all this in a manner closely aligned with real scientific practice. The addition of the structural label to empiricism only covers the minimum required in representations: the different kinds of structural relationship established between mathematical models (*mapping, embedding, etc.*), at different levels; but while necessary, this condition alone is not enough. What else is there in scientific representation? And what really makes it so?

2.5   Use of Models: "Self-Location"

Accepting a theory means "epistemically submitting to its guidance, letting our expectations being moulded by its probabilities regarding observable phenomena." (van Fraassen 1989b.) This is the epistemic dimension of acceptance: we decide to adopt a theory as our expert, and this attitude towards the theory constitutes the perfect definition of acceptance. The image of the "expert" which guides our opinions is, in my opinion, extremely fruitful in that it highlights subjects' attitudes towards the models or hypotheses of science. The idea can be clearly illustrated if we compare our theoretical models to maps, which guide us and enable us to find our bearings. Like maps, theoretical models are partial, are constructed socially in accordance with a series of specific criteria and interests and reflect the concerns and conventions of the era or context in which they are produced.[23] This analogy has also been explored by realist authors such as P. Kitcher (2001) and R. Giere, for whom however, maps are, despite all their constructive elements, partiality and relativity to contexts of use, etc., maps *about something*.

According to Giere's realist interpretation (Giere 2006), in what is, in my opinion, a new clarification of his minimum realist commitments, what makes it possible for us to use maps and models is the fact that they exploit possible similarities between the model and those aspects of the world which are represented. Strictly speaking, however, and here the author agrees with van Fraassen' view, they are not compared with data regarding reality itself, but rather with data models, which implies a level of idealisation and constructivism. The comparison is therefore established between two types of models. There are various constructive and interpretative levels and different fields of research may have different criteria for assessing this meld. Moreover, no one claims that the model itself represents aspects of the world thanks to this relationship of similarity, since no such simple representational relationship exists in science. R. Giere states that: "It is not the model that is doing the representing; it is the scientist using the model who is doing the representing."[24] In other words, they are designed so that some elements of these models may be identified with some characteristics of the real world. This is what makes it possible for us to use models to represent aspects of the world. This is the key; scientists use models to represent aspects of the world in accordance with various purposes, in the same way as we use maps to get our bearings.

However, van Fraassen proposes that we continue to exploit certain characteristics of the map model, providing we trust that it constitutes a good example of the way in which science represents the world. In specific terms, he proposes that we examine the act of using the map itself. Although it is held that its representational power can be testified to by anyone who has ever used a map to get their bearings in

---

[23] In other works I have explored this relationship between models and maps, focusing on the differences between the realists P. Kitcher and R. Ronald Giere and the empiricists H. Longino and van Fraassen. Perdomo (2011).

[24] Giere (2006), 64. The slogan could be, proposes Giere: *No representation without representers.*

unfamiliar territory, it is also true that we need additional information that is not contained in the map itself in order to use it properly. Maps do not include the information "you are here," which we can use to locate ourselves, and even if they do, the act of "self-location" in relation to the arrow which indicates our position is something not included in the map. The act of self-location on or in relation to the map has nothing to do with, or cannot be deduced from, the map's degree of accuracy, nor can it be identified with the contents of the map or with the belief that said map "fits in with" the world, since it does not belong to the semantic field, but rather to the pragmatic one (van Fraassen 1993, 11). The statement that any particular model can be used to represent a specific phenomenon is, according to van Fraassen, an indexical judgement similar to the affirmation that such and such a mark on a map, in relation to which we must locate ourselves, is our actual location. Referring to Kant, van Fraassen states that "the ability to self-attribute a position with respect to the representation is the condition of possibility of use of that representation." (van Fraassen 2008, 257.)

The use of theory to explain, applications to technique, interpretation of data or construction of models are all activities carried out by the scientific community which require a "location" of subjects in relation to the body of knowledge or information in question. To continue with map models, what is characteristic about them, in van Fraassen's opinion, is not their representative function, with all the nuances that can be introduced into said concept, but rather the fact that they constitute useful orientation instruments. From the perspective of empiricism, the model of the map defended by realists, i.e., the model of the map as a constructive representation, albeit, at the end of the day, *representation of*, does not account for the fact that we position ourselves in relation to maps in order to construct them, read them and use them properly. In other words, "self-location" in relation to the map is required for its proper use. van Fraassen again refers to Kant in order to illustrate this point, stating that: "The activity of representation is successful only if the recipients are able to receive that information through their 'viewing' of the representation." (van Fraassen, *Ibid*, 80.) And this is a piece of information not contained in the map or in models; it refers to the relationship established between the model or map, understood as an instrument or artefact, and the interpreting subjects involved in the process of representation, since it is in the act of representation that representations are produced.

We can conceive reality not as a finished structure which must be reproduced from outside, but rather as an open process in which the concept of interpretation gains vital importance. An interpretation which is not retrospective, as in the hermeneutic tradition, but rather prospective, whose aim is precisely to turn reality into intelligible scenarios in which action may be *projected*, in the twofold sense of both planned and pushed forward — a central issue of pragmatism. As a result, we transform reality and interpretative structures should continue adjusting to its movement. We can conceive models as technological artefacts which enable different uses and which can be manipulated and played around with (Morgan and Morrison 1999), we can view them as technologies for research or as fictions which enable us to recreate the feasible or unfeasible possibilities of the behaviour of a phenomenon in a creative and fruitful way. Metaphorical or fictitious licences enable us to explore what would happen to a system of certain characteristics under certain conditions; for this also, computer simulation is, today, a key instrument of model-theoretical research. In this sense, I agree with M. Suárez in recognising the need to develop a more social and pragmatist conception of scientific representation which explores these more dynamic, social and plural aspects, which are characteristic of current scientific and technological practice.

However, at the same time, we can also view the set of "used" and "established" technical, artistic and scientific representations as objects which constitute our world. We can view them as artefacts which become cultural objects to be recreated and interpreted. Let us return to the example of Stein's portrait: some years later, when it *was known* that the picture represented Stein, Picasso is reported to have become angry when he learned that the writer had cut her hair short, although he then thought about it and replied: *"Mais, guand même, tout y est"* (All the same, it is all there). What is all there? We might ask. The system of codes and meanings which make sense of it; the keys to meaning which enable us to locate ourselves in relation to the representation, and which we can reconstruct, understand and interpret; the footprints of our conformations of reality and of our changing interpretations of it throughout history. That's not a realist position, just a way to understand history of science that involves constructivism and contextualism. Science offers us theories which, in addition to being instruments for carrying out tasks in accordance with epistemic or practical objectives, also offer different visions of the world. They are the interpretative coordinates we require to draft the most beautiful cartographies of empirical reality, the ones which will enable us to continue navigating the sea of our intellectual and pragmatic needs. And empiricism offers an adequate vision of this.

References

Brading, K., & Landry, E. (2006). Scientific structuralism: Presentation and representation. *Philosophy of Science, 73*, 571–581.

Cartwright, N. (1999). *The dappled world. A study of the boundaries of science*. Cambridge: Cambridge University Press.

Crombie, A. C. (1994). *Styles of scientific thinking in the European tradition* (3 Vol.). London: Dukworth.

Faerna, A. M. (1996). *Introducción a la teoría pragmatista del conocimiento*. Madrid: S. XXI.

French, S. (2003). A model theoretic account of representation (or, I don't know much about art… but I know it involves isomorphism). *Philosophy of Science, 70*, 1472–1483.

Galison, P. (1996). Constructing modernism: Cultural location of the Aufbau. In R. Giere & A. Richardson (Eds.), *Origins of logical empiricism* (Minnesota studies in the philosophy of science, Vol. XVI, pp. 11–44). Minneapolis: University of Minnesota Press.

Giere, R. (1999). *Science without laws*. Chicago: The University of Chicago Press.

Giere, R. (2006). *Scientific perspectivism*. Chicago: The University of Chicago Press.

Goodman, N. (1976). *Languages of art*. Indianapolis: Hackett.

Hacking, I. (1983). *Representing and intervening*. Cambridge: Cambridge University Press. Spanish Translation: Hacking, I. (1996). *Representar e intervenir* (Sergio Martínez, Trans.). Barcelona: Paidós, UNAM.

Hacking, I. (1991). The self-vindication of the laboratory sciences. In A. Pickering (Ed.), *Science as practice and culture* (pp. 29–64). Chicago: The University of Chicago Press.

James, W. (1897). *The will to believe and other essays in popular philosophy.* Green: Longmans. Spanish Edition: James, W. (2003). *La voluntad de creer. Un debate sobre la ética de la creencia* (Lorena Villamil García, Trans.). Madrid: Ed. Tecnos.

James, W. (1907). *Pragmatism.* Green: Longmans. Spanish Edition: James, W. (1997). *Lecciones de pragmatismo* (Luis Rodríguez Aranda, Trans. Study and notes by Ramón del Castillo). Madrid: Santillana.

Kitcher, P. (2001). *Science, truth and democracy.* New York: Oxford University Press.

Ladyman, J., Bueno, O., Suárez, M., van Fraassen, B. C. (2011). Scientific representation: A long journey from pragmatics to pragmatics. *Metascience, 20*(3), 417–422. Book symposium on Bas C. van Fraassen: *Scientific representation: Paradoxes of perspective.* (Published online, November 2010).

Longino, H. (1990). *Science as social knowledge. Values and objectivity in scientific inquiry.* Princeton: Princeton University Press.

Longino, H. (2002). *The fate of knowledge.* Princeton: Princeton University Press.

Morgan, M., & Morrison, M. (Eds.). (1999). *Models as mediators: Perspectives on natural and social science.* Cambridge: Cambridge University Press.

Moulines, U. (1995). La Filosofía de la Ciencia como disciplina hermenéutica. *Isegoría, 12,* 110–118.

Perdomo, I. (2003). Pragmatismo y empirismo. Acerca de W. James y Bas C. van Fraassen. *Laguna: Revista de Filosofía, 13,* 115–127.

Perdomo, I. (2011). The characterization of epistemology in Philip Kitcher. A critical reflection from new empiricism. In W. J. Gonzalez (Ed.), *Scientific realism and democratic society: The philosophy of Philip Kitcher* (Poznan studies in the philosophy of the sciences and the humanities, pp. 113–138). Amsterdam: Rodopi.

Perdomo, I., & Sánchez, J. (2003). *Hacia un nuevo empirismo.* Madrid: Biblioteca Nueva.

Psillos, S. (2006). *The* structure, the *whole* structure, and nothing *but* the structure? *Philosophy of Science, 73,* 560–570.

Richardson, A. W. (2002). Engineering philosophy of science: American pragmatism and logical empiricism in the 1930. *Philosophy of Science, 69,* 36–47.

Suárez, M. (2004). An inferential conception of scientific representation. *Philosophy of Science, 71,* 767–779.

Suárez, M. (Ed.). (2009). *Fictions in science. Philosophical essays on modeling and idealization.* New York: Routledge.

Suárez, M. (2010). Scientific representation. *Philosophy Compass, 5*(1), 91–101.

Suppe, F. (1974/1977). *The structure of scientific theories* (2nd ed.). Urbana: University of Illinois Press.

van Fraassen, B. (1970). On the extension of Beth's semantics of physical theories. *Philosophy of Science, 37*(3), 325–339.

van Fraassen, B. (1976). To save the phenomena. *The Journal of Philosophy, LXXIII*(18), 623–632.

van Fraassen, B. (1980). *The scientific image.* New York: Oxford University Press.

van Fraassen, B. (1989a). *Laws and symmetry.* New York: Oxford University Press.

van Fraassen, B. (1989b). Probability in physics: An empiricist view. In P. Weingartner & G. Schurz (Eds.), *Philosophy of the natural sciences* (pp. 339–347). Vienna: Hoelder-Pichler-Tempsky.

van Fraassen, B. (1993). From vicious circle to infinite regress, and back again. In D. Hull, M. Forbes, & K. Ohkruhlik (Eds.), *Proceedings of the Philosophy of Science Association, PSA 1992* (Vol. 2, pp. 6–29). Northwestern University Press.

van Fraassen, B. (1994a). Interpretation of science; science as interpretation. In J. Hildevoord (Ed.), *Physics and our view of the world* (pp. 169–187). Cambridge: Cambridge University Press.

van Fraassen, B. (1994b). The world of empiricism. In J. Hilgevoord (Ed.), *Physics and our view of the world* (pp. 114–134). Cambridge: Cambridge University Press.

van Fraassen, B. (1994c). Gideon Rosen on constructive empiricism. *Philosophical Studies, 74,* 179–192.

van Fraassen, B. (1997). Structure and perspective: Philosophical perplexity and paradox. In M. L. Dalla Chiara et al. (Eds.), *Logic and scientific methods* (pp. 511–530). Netherlands: Kluwer Academic Publishers.

van Fraassen, B. (2002). *The empirical stance.* New Haven/London: Yale University Press.

van Fraassen, B. (2008). *Scientific representation: Paradoxes of perspective.* New York: Oxford University Press.

van Fraassen, B., & Sigman, J. (1993). Interpretation in science and in the arts. In G. Levine (Ed.), *Realism and representation* (pp. 73–99). Madison: University of Wisconsin Press.

Chapter 3
Models and Phenomena: Bas van Fraassen's Empiricist Structuralism

Valeriano Iranzo

Abstract Bas van Fraassen's recent endorsement of *empiricist structuralism* is based on a particular approach to representation. He sharply distinguishes between what makes a scientific model *M* a successful representation of its target *T* from what makes *M* a representation of *T* and not of some other different target T'. van Fraassen maintains that *embedment* (i.e.: a particular sort of isomorphism which relates structures) gives the answer to the first question while the user's decision to employ model *M* to represent *T* accounts for the representational link. After discussing the rationale for this approach, I defend that indexical constraints like those favoured by van Fraassen cannot be the last word concerning what makes a scientific model a representation of something in particular. Rather, I argue that (i) the representational role of models — at least of scientific models — is inextricably related to their ability to convey some knowledge about their purported target, and (ii) this is an effective constraint on the user's decisions. Both claims cast some doubt on the aforementioned distinction insofar as not only success in representation, but also the existence of a representational relation, is rooted in our knowledge about the target.

Keywords Models • Empiricist structuralism • Scientific representation • Isomorphism • van Fraassen

---

This work was funded by the Spanish Ministry of Science and Innovation (research project FFI2008-01169).

V. Iranzo (✉)
Faculty of Philosophy and Education Sciences, University of Valencia,
Blasco Ibáñez Av. 30, 46010 Valencia, Spain
e-mail: valeriano.iranzo@uv.es

W.J. Gonzalez (ed.), *Bas van Fraassen's Approach to Representation and Models in Science*, Synthese Library 368, DOI 10.1007/978-94-007-7838-2_3,
© Springer Science+Business Media Dordrecht 2014

3.1   Introduction: Models as Representational Devices

Modeling plays an important role in current scientific practice. Nonetheless, it is not easy to give a clear-cut answer to the question of what is a scientific model since very different sorts of entities may be so considered. Compare wood models of molecules and their contemporary surrogates, that is, three-dimensional computer generated images. Now think of some other examples as the ideal gas model, Maxwell's ether model, Bohr's model of the atom, the Fisher-Wright model in population genetics, the Hardy-Weinberg equilibrium rule. Modeling is related to such heterogeneous entities as wood pieces, images on a computer screen, equations, … It should also be noticed that those things represented by models — the *target* of the model — can be as diverse at least as those objects that do the representational work, that is, models themselves. Heterogeneity is, then, intrinsic to modeling — and also to representation.

Is there any common feature to all these different entities in virtue of which they can be considered as models? The usual reply is *representational ability*.[1] Representations may be linguistic (descriptions), mathematical, pictorial, three-dimensional, …, and models are mainly used, then, to represent objects, systems, processes, sets of data, … They are representational devices but, since not all representations are models — a portrait, for instance, is a pictorial representation, but not a scientific model — modeling is a peculiar way of representing things frequently employed in science.[2]

Representations may be better or worse depending on the aims pursued. Think of a car engine. What may be forcefully good for teaching people who tries to get a driving licence, is probably very sketchy for engineers working on improving its fuel efficiency. Anyway, in both cases something is represented. If the model — a draw, a graph, …, in this case — did not represent a car engine, we could hardly get some *knowledge* of car engines by means of this model. In fact, those who claim that representing is fundamental in modeling assume this link between representation and knowledge. In favour of this standpoint it is perhaps worth adding that scientists look for representations with some epistemic import about the target represented. That is a crucial difference to what happens with representation in the fine arts, for instance.[3]

What is the relation between scientific theories and those representational devices called models? The classical view on scientific theories defends that they are linguistic-propositional entities. The goal of axiomatization is to show that the theoretical principles are linked to empirical data by virtue of correspondence rules that partially define theoretical terms by means of observational ones. In keeping

---

[1] Different interpretations of this basic claim can be found in Cartwright 1999; Giere 1988, 2004; Hughes 1997; Morrison 2009; Suppe 1989; van Fraassen 1980, 2008.

[2] For a discrepant view, see Knuuttila 2005, where it is argued that in order to understand modeling in scientific practice, the focus on representation is unnecessary limiting.

[3] See, however, Callender and Cohen 2006.

with this syntacticist account, models were understood as convenient implementations for theories given that theoretical principles are usually very abstract. When exposing the theory models are helpful to understand what the theory really says; they are also useful for ascertaining how the theory can be applied in particular situations; they may suggest unconsidered possibilities for extending the theory to new domains, … The classical view not only accepts that models play a *heuristic role* in science. It may even accept that models decisively increase the explanatory power of theoretical principles in respect of experimental laws (see Nagel 1961, chap. 6). In addition to this, modeling is seen as a non-autonomous task insofar as the rationale for doing it is the development and application of a given theoretical framework.

In contrast to the foregoing, the semantic view of theories — whose advocates are Patrick Suppes, Joseph Sneed, Bas van Fraassen, Frederick Suppe, and Ronald Giere among many others — emphasizes the importance of modeling in science. For semanticists a theory is not a set of statements but a set of models. In van Fraassen's words:

> To present a theory, we define the class of its models directly, without paying any attention to questions of axiomatizability, in any special language, however relevant or simple or logically interesting that might be. And if the theory as such is to be identified with anything at all—if theories are to be reified—then a theory should be identified with its class of models.[4]

Consequently, models are not just heuristic tools subsidiary to theories: if a theory is no more than a collection of abstract objects, there is no qualitative difference between the roles assigned to theories and those assigned to models. Furthermore, to the extent that theories attain any sort of epistemic values like understanding, knowledge, empirical adequacy, truth, …, so do models.[5]

In the following I will focus on the semantic approach, particularly on the structuralist account of models.

3.2 Similarity

Even though heterogeneity is intrinsic to representation it makes sense to ask what makes that *A* (the model) represents *B* (the target)?

There are two main options here. "Informational" views emphasize objective relations between models and their target systems, while "functional" views focus on cognitive activities related to these targets enhanced by models — like inference or interpretation (Chakravartty 2010). Since van Fraassen's account of representation is "informational," functional views will be put aside here.[6]

---

[4] van Fraassen 1989, p. 222. This is a strong formulation that tries to keep distance from a "partially linguistic" view on models. See below, footnote 12.

[5] Nancy Cartwright and Margaret Morrison, among others, have insisted that modeling is an activity autonomously pursued in respect of theories. See Cartwright 1999; Morrison and Morgan 1999 and Morrison 1999.

[6] Functional accounts of scientific representation can be found in Suárez 2004 and Contessa 2007.

What sort of relation could be expected between the model and its target? Since models are not linguistic entities, truth does not seem an appropriate candidate here. To consider that *A* is true of *B*, i.e., that *A* is a true description of *B*, would be a sort of categorical mistake. In addition to this, models usually involve idealizations, sometimes obvious distortions, of the target system — the ideal-gas model may be a good example here. Hence many scientific models are literally false taken as a whole, although they are successful in representing their target.[7]

An appealing alternative to truth is *similarity*. Some models at least are similar to their target systems in some respects. Wood models of molecules resemble real molecules, but only in some respects — the latter are much smaller that their wooden counterparts. This is the option developed by Ronald Giere in his particular interpretation of the semantic approach to scientific theories. However, there are many respects in which two different things may be similar/different. Articulating general principles to discern the relevant respects of similarity involved when the comparison is made between an abstract model and its target seems a lost cause. Hence Giere maintained that similarity is unanalyzable: it cannot be explicated in terms of any other more basic relation.[8]

More ambitious approaches to define similarity try to cash it out in formal terms. According to a "purely structuralist" view of models, the one favoured by van Fraassen, "models are mathematical entities, so all they have is structure […]."[9] A structure *S* is a composite entity consisting of a non-empty set of individuals — the domain *D* of the structure *S* — a non-empty set *R* of relations on *D*, and a set of operations (which may be empty) on *D*, that is:

$$S = < a_1,....,a_n, R_1,...R_n, o_1,...,o_n >$$

From this standpoint, modeling basically consists in elaborating *structures*, and the representational ability of scientific models essentially depends on a special sort of similarity, i.e., *structural similarity*.

A particular version of structural similarity is *structural isomorphism*. Intuitively, two systems S_1 y S_2 are similar, from a structural point of view, if a correspondence between the elements (namely, individuals, relations and operations) of both taken

---

[7] For contrasting opinions on the alleged fictional status of scientific models, see Suárez 2009, and Iranzo 2011.

[8] Giere 1988, p. 80. Giere's views have evolved to an intentional conception of similarity and representation as we will see below.

[9] van Fraassen 1997, p. 528. For a "partially linguistic" account of models, they are not bare mathematical structures, but a sort of mixed compound: structure plus linguistic interpretation. Although both alternatives — pure structuralism and partial linguisticism — have room within the semantic approach to scientific theories, structuralists like van Fraassen forcefully insist that their option is radically different from the classical syntacticist view of theories. The technical question at issue is the possibility of a first-order axiomatization of the class of models whereby the theory is identified. Pure structuralism rejects this possibility. See Da Costa and French 2003, chap. 2, and Suárez 2005, par. 3.

one by one may be established. In that case it is said that S_1 and S_2 are isomorphic.[10] Since isomorphism is a function that establishes a one-to-one mapping, the cardinality of their respective domains must be the same and the number of relations defined over them is also identical. However, objects and relations found in S_1 and S_2 could be different. Only formal properties of these elements are preserved — for instance, an equivalence relation over a domain of cardinality n is mapped onto another equivalence relation over a domain of equal cardinality.

Isomorphic systems could be seen, then, as different interpretations of the same underlying structure. But we cannot simply say that A represents B *iff* there is a structural isomorphism between A and B. Structural isomorphism is a symmetrical relation while representation is not (Suárez 2003). The Lotka-Volterra model of predator — prey interactions, for instance, represents interspecific competition in a particular population. It is not very realistic — it does not consider any competition among prey or predators, for instance — but it represents populations, insofar as there is some structural similarity between the model and natural populations. Yet populations do not represent the model. Then, a further condition is demanded in order to circumvent symmetry.

Since models are *used* by scientists to represent processes, phenomena, …, and not the other way round, a straightforward possibility is to include the user's intentions as an additional restrictive condition. Thus Giere has modified his previous approach in favour of an "intentional conception" of scientific representation that distinguishes the representation, as a result or a product, from the activity performed by the agent (Giere 2010). Given that "scientific practices of representing the world are fundamentally pragmatic," in order to understand them we should focus on the *activity of representing* instead of *representation*. Consequently, the relevant relation is not a simple dyadic relation like "X represents W" but a more complex one: "S uses X to represent W with purposes P." (Giere 2004, 743.) Giere still maintains that similarity is the desired relation between models and the world, although now he insists that qualifications in respects and degrees of similarity must be *intentionally* qualified.

I will not pause on Giere's interesting proposal. I introduce it here to show that the pragmatic-contextual dimension of representation is a matter of concern for both formal and non-formal accounts of representation developed within the semantic tradition. The general idea is that the existence of a representational link, which goes just in one direction, cannot be fully explicated by focusing only on the relation between the model and the target system. Agents also play a substantial role here.

van Fraassen introduces pragmatic considerations on modelling by distinguishing between the adequacy of A as a representation of B, on one side, and the condition of possibility of using A for representing B:

> "On the semantic view, a theory offers us a large range of models…. If a theory is advocated then the claim made is that these models can be used to represent the phenomena, and to represent them accurately. A model (can be used) to represent a given phenomenon

---

[10] In what follows I will put aside operations since they can be reduced to relations: an operation taking n arguments is equivalent to a $n+1$ place relation.

accurately only if it has a substructure isomorphic to that phenomenon. (That structural relationship to the phenomenon is of course not what makes it a representation, but what makes it accurate: it is its role in *use* that bestows the representational role) (van Fraassen 2008, 309).

We should not confuse, then, the accuracy of a model in respect of a particular target from its representational function. I will deal with van Fraassen's empiricist account of the accuracy of theoretical models in the next section. The pragmatic factor involved in representation will be discussed in Sect. 3.4.

3.3 Structural Empiricism

According to the quoted paragraph in the previous section van Fraassen sees models as tools by means of which *theories represent phenomena*:

> The behaviour of pendulums and bouncing springs was well-known by Newton's time, but he *represented both* as systems subject to a force varying directly with the distance from a midpoint. Today we would say that Newton's theory provides models satisfying $F = -kx$, and *these models can be used to represent such phenomena* (my emphasis) (van Fraassen 2010, 511).

In fact, theoretical models "are provided in the first instance to fit observed and observable phenomena." (van Fraassen 2008, 168.)

But scientific modeling is a multi-faceted task. The information given by experiments looks fragmentary and disparate in the light of theoretical models unless that information is systematized by further low-level models. Thus, in addition to theoretical models van Fraassen distinguishes between *data models* and *surface models* (van Fraassen 2008, 166 and ff).

Data models are in close contact with raw data. They represent the outcome of an operation — the value for the patient's temperature, for instance — through a number, a pair of numbers (e.g., mean and standard deviation), a graph with the relative frequencies found in more complex examples that involve multiple operations in various times/locations/patients…[11] These refined data are highly discrete and not well-suited yet to theoretical models. The graph must be abstracted into a mathematically idealized form by smoothing over discrepancies, by assuming non-discrepant values in cases where the measurement operation did not take place, … Relative frequencies give way to density functions. As a result a continuous range of values is obtained. Surface models are precisely those idealized representations — actually they are mathematical structures — necessary for confronting theoretical models with data.

How can be ascertained whether a particular theoretical model M_t is an adequate representation of some phenomena? The requirement is that "the data or surface models must ideally be isomorphically embeddable in theoretical

---

[11] It should be noticed that for van Fraassen measuring is "a practical form of representation presupposing a prior theoretical representation," van Fraassen 2010, pp. 512–513. Chapters 5 to 8 of *Scientific representation* are devoted to this issue.

models." (van Fraassen 2008, 168.) van Fraassen resorts to a technical notion, i.e., *embedding*, to explain how theoretical models are successful in representing phenomena. Let me pause on this. An *empirical substructure* of a theoretical model is strictly a subset of the domain of the latter. The relations included in the substructure are the restrictions of the relations of the theoretical model to the smaller domain of objects, and only those objects, considered in the substructure. Adequate representation demands isomorphism just between empirical substructures of theoretical models and "phenomenological" structures — data models and surface models. It is clear that *embedding* is a formal constraint less demanding that isomorphism *tout court*, insofar as the one-to-one mapping is not required for the whole domain of the theoretical model.[12]

Now, if a substructure of M_t is structurally isomorphic to a data model of some phenomenon F, then M_t does represent successfully F. That is what *isomorphic embedding* consists in. Thus, a theoretical model that satisfies the Newtonian Law of Gravitation represents the planetary motion in the solar system. Kepler elaborated a surface model from the observations made by Tycho Brahe about these phenomena. Since the structure of Kepler's model is isomorphic to a substructure of the Newtonian theoretical model, the latter accurately represents the planetary motion in the solar system ... provided that astronomers effectively *use* Brahe's data model to represent *those* phenomena (the particular planetary movements recorded by him). But before discussing how this pragmatic factor comes on stage, some comments are in order here.[13]

According to van Fraassen's empiricist standpoint, science aims at *empirical adequacy* and modeling is mainly intended both to fit and to represent phenomena. Consequently, he equates accuracy in representation to an empiricist version of structural isomorphism, i.e., embedding, so that "if we try to check a claim of adequacy, we will compare one representation or description with another — namely, the theoretical model and the data model." (van Fraassen 2006, 545.)

It is important to realize, however, that for van Fraassen structuralism is not just a methodological thesis about what scientific theories are. Rather, it must be also understood as an epistemic view which asserts that "all we know is structure." The core claims of this *empiricist structuralism* are:

1. Science represents empirical phenomena as embeddable in certain abstract structures (theoretical models).
2. Those abstract structures are describable only up to structural isomorphism.[14]

---

[12] Partial isomorphism and homomorphism are some other criteria less demanding than structural isomorphism. See Da Costa and French 2003, chap. 3, for a defense of partial isomorphism; Mundy 1986, and Bartels 2006, favour homomorphism.

[13] More than thirty years ago van Fraassen claimed that *empirical adequacy* for theoretical models involves embedding (see his *The scientific image*, p. 45 and p. 64). In *Scientific representation*, however, he introduces substantial changes on his old view. Firstly, he emphasizes that embedding is not enough to account for the models' ability in representing phenomena since the agents as user plays a crucial role. Secondly, phenomena themselves have no structure in contrast to models.

[14] van Fraassen 2008, p. 238. There is still an *ontic* version of structuralism according to which "all that there is, is structure." See Ladyman 1998.

Data models and surface models are indispensable mediators between those theoretical models and phenomena. The aim of scientific modeling taken as a whole is "saving the phenomena" and this could not be possible unless those *observed and observable phenomena* were represented in one way or another. But if isomorphism takes place between structures and it is necessary for successful representation, then it seems that models could not represent phenomena unless phenomena themselves had any definite structure that good models somehow apprehended.

van Fraassen acknowledges that he was wrong on this point in *The Scientific Image*, where isomorphism was established between a substructure of the model-theoretic and the phenomena (van Fraassen 2008, 386). Now he discards a realist interpretation of structuralism that locates structures in the extra-scientific world since structural empiricism is "a view not of what nature is like but of what science is."[15]

In addition to this, van Fraassen warns us not to conflate appearances and phenomena: "*Phenomena* are observable entities (objects, events, processes, …) of any sort, *appearances* are the contents of measurements outcomes."[16] Appearances are the manifestations of phenomena recorded by our instruments for detection and measurement. The things actually observed and recorded are appearances: "…by definition, we never do see beyond the appearances…!" (van Fraassen 2008, 99). Both surface models and data models summarize appearances. Now if empirical adequacy requires structural isomorphism between them and substructures of theoretical models, it seems that theoretical models are empirically adequate only in respect of those appearances represented in data models. In that case theoretical models would successfully represent appearances, i.e., experimental data, but what about phenomena? How can they be saved by theoretical models? van Fraassen's reply to this challenge — the "Loss of Reality Objection" as he labels it — invokes the pragmatic constraints on representation as we will see in the next section.

3.4   Pragmatic Tautologies

van Fraassen insists that phenomena themselves do not dictate which structures represent them. Consequently, the representational link between the data model and its target cannot be specified in a non-indexical way. It is the user's model who *decides* that *this* data model represents *this* phenomenon.

Suppose that S obtains a graph after a lengthy and careful empirical research. S takes it to represent a phenomenon, for instance, the population growth of red

---

[15] van Fraassen 2008, p. 239. This is a radical departure from realist — a synonym for "metaphysical," according to van Fraassen — interpretations of structuralism, such as Ladyman and French's ontic structuralism (see above, footnote 14).

[16] van Fraassen 2008, p. 283. This proposal is parallel to the distinction between phenomena and data drawn in an influential paper: Bogen and Woodward 1988. However, Bogen and Woodward consider phenomena as "not observable in any interesting sense of that term."

deer in a National Park for several years. Once S establishes a theoretical generalization that fits with the graph, that is, with the data model, there is no difference *for her* between:

(a) M_t is empirically adequate to phenomena F.
(b) M_t is empirically adequate to phenomena F as represented by S.

The truth conditions of (a) and (b) are not the same, certainly. But "(a)=(b)" is a "pragmatic tautology," according to van Fraassen, since S cannot deny (a) and assert (b) at once. That would be a *pragmatic contradiction*, that is, a logically contingent statement that cannot be asserted. This divergence between the semantic (truth/falsity) and the pragmatic status of a statement (assertability/deniability) is salient when the statement contains an indexical element. Therefore, even though "(a)=(b)" may be false, "(a)≠(b)" cannot be asserted by S. In particular, "in a context in which a given model is someone's presentation of a phenomenon, there is *for that person* no difference between the question *whether a theory fits that representation* and the question *whether that theory fits the phenomenon*." (van Fraassen 2008, 260.) Given that empirical adequacy is restricted to comparison of representations (models), S has no way to ascertain the empirical adequacy of M_t concerning plain F. The only possibility open to her is to ascertain it concerning F as represented by this or that surface/data model. So, there is no (pragmatic) difference for her. van Fraassen concludes then that the "Loss of Reality Objection" is dissolved.

It is worthwhile to notice that the pragmatic impossibility for S to distinguish between (a) and (b) does not entail that S could not discover that her representation of F is severely misguided. Further developments in measurements could suggest that a particular data-model M_d does not accurately represent an observable phenomenon F. In that case, S would be compelled to revise her previous assessments on the empirical adequacy of M_t, a theoretical model structurally isomorphic to M_d. Now, let us suppose that S replaces M_d for a different representation of F, i.e., $M_d{}^*$, a data model that fits better with those new measurement outcomes. After elaborating a new theoretical model, $M_t{}^*$, in which $M_d{}^*$ can be embedded, S's ability for assessing empirical adequacy is the same as before. Again, "$M_t{}^*$ is empirically adequate to F" is pragmatically indistinguishable for S from "$M_t{}^*$ is empirically adequate to F as represented by $M_d{}^*$." If pragmatic impossibility prevented us from discerning which models are empirically adequate and which ones are not, empiricism would collapse.

On the other side, disagreement between theoretical models, regarding the content which cannot be encapsulated in any empirical substructure, is no matter of concern for an empiricist like van Fraassen. Two incompatible theoretical models can be empirically adequate to F insofar as each one contains a substructure isomorphic to a model of data which represents F. We cannot *believe* that both models are correct, sure, but van Fraassen's well-known point is that beliefs about their truth/falsity are fully dispensable in scientific practice.[17] Theoretical models can do their work

---

[17] "…the basic aim [of science] — equivalently, the base-line criterion of success — is empirical adequacy rather than overall truth, and that acceptance of a scientific theory has a pragmatic

properly — they can be used as tools for prediction, for instance — notwithstanding *S*'s agnosticism about them. *S*'s beliefs on how is the world in those aspects that exceed empirical adequacy are completely irrelevant regarding scientific practice. So, both models could be *accepted* by *S*, even though she did not believe any of them. In sum, theoretical disagreements can perfectly coexist with empirical adequacy.

Now we have an answer to why this particular data model represents a phenomenon *F* – red deer population growth in a National Park – and does not represent a different one – the star formation rate in a galaxy, for instance. According to van Fraassen, we must look at the way data models are used: *S* uses that model to represent the former phenomenon, but not to represent the latter. The user is crucial in the representational relation.

To cut a long story short, M_d represents this phenomenon *F* because a human agent uses it to represent *F*. Granted that M_d represents *F*, then, if embedding is also satisfied, the theoretical model M_t is an accurate representation of *F*. If isomorphism fails, M_t would not be a good representation of that very phenomenon *F* which M_d effectively represents insofar as it is used to do that. Furthermore, the user cannot assert that M_t is empirically adequate to *F* and not to *F* as represented by M_d. She can neither assert that M_t is empirically adequate to *F* as represented by M_d and not to *F*. From the foregoing van Fraassen concludes that scientific theories are not about the world as represented by us in any idealistic sense (van Fraassen 2006, footnote 11).

3.5   Representing and Knowing

Recall that appearances are perspectival manifestations of phenomena accessible to us: "… the measurement outcome shows not how the phenomena *are*, but how they *look*." (van Fraassen 2008, 290.) Besides, the definition of empirical adequacy as structural embedding implies that the empirical adequacy of theoretical models obtains in respect of appearances, provided that phenomena themselves cannot be structurally described. If knowledge is, according to empiricist structuralism, knowledge about structures it can hardly be explained how we get knowledge of phenomena since they are something beyond the appearances. The pragmatic tautology invoked by van Fraassen highlights a discursive constraint which should not be overlooked, certainly, but it does not dispel this image of an agent who is trapped in a world of appearances.

But I will not press the point here. I will focus, rather, on the sufficiency of the pragmatic constraint to account for the representational link: if the appropriateness

---

dimension (to guide action and research) but need involve no more belief than that the theory is empirically adequate." van Fraassen 2008, p. 3. This idea was van Fraassen's motto in *The scientific image*. For a criticism on the alleged redundancy of belief in scientific practice, see Iranzo 2002.

of a data model to represent F depends just on "our selective attention to it and our decisions to represent them in certain ways and to a certain extent," (van Fraassen 2008, 254) it seems that any data model could represent any phenomenon if S's decision were so and so. That conclusion would be a reductio for any account of scientific representation.

It can be added that the individual decision taken by S is not necessarily arbitrary. Models of data must be elaborated according to standard methodological procedures operating in scientific research. Avoiding biased samples, for instance, is a good policy for arriving at reliable estimations of the real value of a parameter. But to minimize the subjectivity of individual decisions by appealing to consensus among scientists, to collective decision, etc., misses the point, in my opinion.

Representational practices pursue very diverse aims in different human activities. But using the symbol "$" for representing money may be a very useful convention seems very different from using a data model to represent a phenomenon. There is nothing in money which bestows the representational role on this symbol instead some other one like "&." Besides, by representing money in such way we do not gain any knowledge of it. In contrast, I take it that the most peculiar scientific representational practices are mainly guided by epistemic values. Putting the matter in other words, a scientific model provides some knowledge of the target insofar as that model is successful in representing it, leaving aside if it is a more or less accurate representation of this target. If a model does not give us any knowledge, understanding, ..., about its alleged target, it can hardly be a representation of it. Models of data are no exception here: its representational role is inextricably related to its ability to convey some knowledge about its purported target.

For the sake of the argument I will take for granted an empiricist standpoint — scientific knowledge is only about the observable dimension of the world — and I will also assume that the aforementioned gap between appearances and phenomena is bridged. Then, it could be said that models of data provide knowledge of those phenomena represented by them.

¿How could we get some knowledge of F through a structural representation of it? If F contained the same structure we find in M_d, perhaps it could be defended that representing F in this format conveys some knowledge of it, at least about its structural features. It should be recalled here that van Fraassen explicitly rejects this option since he is not willing to endorse a realist standpoint on universals (and particularly on structures) (van Fraassen 2008, 247). But even though data models do not completely exhaust F, the phenomenon somehow constrains which models of data do represent it and which other do not. Otherwise, any model of data could represent any phenomena, and that would be unpalatable for a theory of scientific representation, as I pointed out above.

The particular data model employed by S for representing F effectively represents F only if it provides some knowledge about F, in contrast to some other models which would be completely idle from this epistemic point of view and which would not be considered representations of F precisely for this reason. So, S's preference for a particular data model on this occasion is conditional on its epistemic value, and the latter, in its turn, does not depend only on how the model is used by *S*, but *on the extent that this model reveals something about the nature of this particular phenomenon.*[18]

The foregoing suggests, on my opinion, that *S*'s decision for using *this* data model M_d for representing *this* phenomenon *F* on *this* occasion cannot determine by itself that M_d represents *F*. Equating "A represents B" — as different from "A accurately/roughly/… represents B" — to "A is used (by *S*) for representing B" overlooks the link between representing and knowledge in science. It is difficult to explain why this data model gives us some knowledge about its target insofar as it represents it just by pointing at the brute fact that it is used for representing it.

Perhaps we should accept that representation is not the only way to gain access to phenomena. In this vein, M. Ghins has claimed that the reliability of a data model "relies on some basic truths about real observational facts, with respect to which our construction of a representation is, so to speak, parasitic." (Ghins 2010, 533.) Statements like 'this is a gas' or 'this gas is hotter than this other gas' are observational, insofar as their truth is ascertained on the basis of direct observation. I will add that data models assume that phenomena are categorized in some particular way. They appropriately represent phenomena insofar as the categorization is fine. The "basic truths" would not state a representative relationship, in contrast to measurements of a property. They will attribute properties to things as we usually do in many of our daily assertions.

van Fraassen, notwithstanding, claims that our decisions to represent phenomena in certain ways and to a certain extent assume that we have pre-scientific ways of *describing* them, (van Fraassen 2008, 387, footnote 20) but he does not go into further details about the role played by these descriptions in scientific representations. The point I want to emphasize here, however, is that the truth of those descriptions may be relevant after all for assessing whether the phenomenon is represented by a particular model. Theoretical models would be empirically adequate in respect of those aspects represented by "low-level" scientific models (data models and surface models). But successful representation of phenomena demands that the "basic truths" assumed by the data model are true.[19]

When truth goes on stage, there is some risk of introducing an excess of metaphysical baggage for an empiricist feeling. van Fraassen rejects truth as correspondence, since it assumes "a user-independent relation between words and things that determines whether a sentence is true or false," (van Fraassen 2008, 252) but he also admits "a common sense realism in which reference to observable phenomena is unproblematic." (van Fraassen 2008, 3.) A detailed discussion on truth is beyond the scope of this paper, of course, but it should be added that the semantic value of many assertions about observable phenomena does not seem problematic either.

---

[18] On the insufficiency of the structuralist account of knowledge see Psillos 2006, pp. 566 and ff., where it is argued that identification of structures depends on knowledge about non-structural properties of the object that "fill" the structures.

[19] The idea that a structure can represent a target system only with respect to a certain description of it that is true is argued from a different perspective in Frigg 2006, 55 and ff.

So, in principle, common sense does not seem disturbed just by talking about such sort of truths. On my view, in addition to the indexical constraints highlighted by van Fraassen, truth is also required to understand how science is successful when representing phenomena. It could be seen as a further condition of possibility for scientific representation, in addition to the user's role. Consequently, truth is somehow more fundamental than representation.

References

Bartels, A. (2006). Defending the structural concept of representation. *Theoria, 55*, 7–19.

Bogen, J., & Woodward, J. (1988). Saving the phenomena. *Philosophical Review, 97*, 303–352.

Callender, C., & Cohen, J. (2006). There is no special problem about scientific representation. *Theoria, 55*, 67–85.

Cartwright, N. (1999). *The dappled world. A study of the boundaries of science*. Cambridge: Cambridge University Press.

Chakravartty, A. (2010). Informational versus functional theories of scientific representation. *Synthese, 172*(2), 197–213.

Contessa, G. (2007). Representation, interpretation and surrogate reasoning. *Philosophy of Science, 74*, 48–68.

Da Costa, N. C. A., & French, S. (2003). *Science and partial truth: A unitary approach to models and scientific reasoning*. Oxford: Oxford University Press.

Frigg, R. (2006). Scientific representation and the semantic view of theories. *Theoria, 21*, 49–65.

Ghins, M. (2010). Bas van Fraassen on scientific representation. *Analysis, 70*, 524–536.

Giere, R. (1988). *Explaining science*. Chicago: The University of Chicago Press.

Giere, R. (2004). How models are used to represent reality. *Philosophy of Science, 71*, S742–S752.

Giere, R. (2010). An agent-based conception of models and scientific representation. *Synthese, 172*(2), 269–281.

Hughes, R. I. G. (1997). Models and representation. *Philosophy of Science, 64*, 325–336.

Iranzo, V. (2002). Constructive empiricism and scientific practice: A case study. *Theoria, 17*, 335–357.

Iranzo, V. (2011). Ciencia, modelos, ¿ficciones? *Teorema, 20*, 151–173.

Knuuttila, T. (2005). Models, representations and mediation. *Philosophy of Science, 72*, 1260–1271.

Ladyman, J. (1998). What is structural realism? *Studies in History and Philosophy of Science, 29*, 409–424.

Morrison, M. (1999). Models as autonomous agents. In M. Morrison & M. S. Morgan (Eds.), *Models as mediators: Perspectives on natural and social sciences* (pp. 38–65). Cambridge: Cambridge University Press.

Morrison, M. (2009). Fictions, representations and reality. In M. Suárez (Ed.), *Fictions in science* (pp. 110–135). Abingdon: Routledge.

Morrison, M., & Morgan, M. S. (1999). Models as mediating instruments. In M. Morrison & M. S. Morgan (Eds.), *Models as mediators: Perspectives on natural and social sciences* (pp. 10–37). Cambridge: Cambridge University Press.

Mundy, B. (1986). On the general theory of meaningful representation. *Synthese, 67*, 391–437.

Nagel, E. (1961). *The structure of science*. New York: Harcourt Brac.

Psillos, S. (2006). The structure, the whole structure, and nothing but the structure. *Philosophy of Science, 73*, 560–570.

Suárez, M. (2003). Scientific representation: Against similarity and isomorphism. *International Studies in the Philosophy of Science, 17*, 225–244.

Suárez, M. (2004). An inferential conception of scientific representation. *Philosophy of Science, 71*, S767–S779.

Suárez, M. (2005). The semantic view empirical adequacy and application. *Crítica. Revista Hispanoamericana de Filosofía, 37*, 29–63.

Suárez, M. (Ed.). (2009). *Fictions in science*. Abingdon: Routledge.

Suppe, F. (1989). *The semantic conception of theories and scientific realism*. Chicago: University of Chicago Press.

van Fraassen, B. (1980). *The scientific image*. Oxford: Clarendon.

van Fraassen, B. (1989). *Laws and symmetry*. Oxford: Oxford University Press.

van Fraassen, B. (1997). Structure and perspective: Philosophical perplexity and paradox. In M. L. Dalla Chiara et al. (Eds.), *Logic and scientific methods* (pp. 511–530). Dordrecht: Kluwer Academic Publishers.

van Fraassen, B. (2006). Representation: The problem for structuralism. *Philosophy of Science, 73*, 536–547.

van Fraassen, B. (2008). *Scientific representation*. Oxford: Oxford University Press.

van Fraassen, B. (2010). Scientific representation: Paradoxes of perspective – Summary. *Analysis, 70*, 511–514.

Part II
Models and Representations

Chapter 4
The Criterion of Empirical Grounding in the Sciences

Bas C. van Fraassen

Abstract A scientific theory offers models for the phenomena in its domain; these models involve theoretical quantities of various sorts, and a model's structure is the set of relations it imposes on these quantities. There is an important, indeed fundamental, demand in scientific practice that those quantities be clearly and feasibly related to measurement procedures. The scientific episodes examined include Galileo's measurement of the force of the vacuum, Atwood's machine designed to measure Newtonian theoretical quantities, Michelson and Morley on Fresnel's hypothesis for light aberration, and time-of-flight measurement in quantum mechanics. The fundamental demand for *empirical grounding* is then given a precise formulation following this scrutiny of crucial junctures where the role of theory in measurement came clearly to light.

Keywords Scientific models • Measurement • Theory-dependence of measurement • Theoretical quantities • Empirical grounding

4.1   The Interplay of Theory, Model, and Measurement

The relationship between theory and phenomena involves an interplay of theory, modeling, and experiment during which both the identification of parameters and the physical operations suitable for measuring them are determined. Recognizing

Research for this paper was supported by NSF grant SES-1026183. A short version of this paper was presented at the Philosophy of Science Association Conference 2010 with the title "Modeling and Measurement: The Criterion of Empirical Grounding."

B.C. van Fraassen (✉)
Department of Philosophy, San Francisco State University,
w/n, 94132 San Francisco, CA, USA
e-mail: fraassen@sfsu.edu

this interplay has sometimes been suspected of threatening the objectivity of science. Peter Kosso, for example, calls for a "declaration of independence" between theory and experiment:

> Insofar as observation is theory relative in the sense that theory influences not only what observations are to be made but also what those observations mean, the accountability of scientific claims is an internal affair and the reliability of science is self-proclaimed. So why should we believe in science?
>
> [B]y now many philosophers have conceded to a certain amount of theory-dependence in observation and its role as an objective standard is threatened. (Kosso 1989, 245–246).

Alan Chalmers (2003) aptly describes Kosso's requirement as a preventative measure against *theoretical nepotism*.

But if that prevention were truly thorough, would it not leave the experimenter theoretically illiterate? That such theoretical neutrality is just not feasible is a theme familiar from Thomas Kuhn. In "The function of measurement," Kuhn displays the pitfalls in the idea of pure data generating theory as a simplistic picture of a "theory machine:" the data are fed in, a crank is turned, and a confirmation or disconfirmation is disgorged.

Kuhn addresses simultaneously the cliché that the theory can be back-inferred from those data, and the companion that what counts as experiment, measurement, or data is independent of what the theory is or says, that it is neutral between theories. But these clichés drive a quite common conception of the scientific enterprise as similar to a Sherlock Holmes-like investigation to settle, with autonomous data, the question of truth or falsity of the detective's hypotheses. That conception, as Kosso's cautions indicate, is threatened by the realization of a constant interplay between the construction of models, formulation of hypotheses, designs for experimental and measurement apparatus. At the same time, that interplay clearly succeeds in bringing information about the studied phenomena to light.

What is needed to counteract both the threat of theoretical nepotism undermining scientific inquiry and the simplistic common conception that it threatens is a thorough scrutiny of the normative requirements that govern such inquiry. That means first of all investigating measurement as it is proposed, designed, and carried out by scientists, to elicit the actual role of theory or modeling in measurement.

In conclusion I will then locate this view of measurement in the larger picture of science subject to the demand of empirical grounding. This will provide a corrective to the very relevance of those worries about "theory-infection," without undermining the empirical character of the sciences. To the extent that they presume or presuppose independence between theory and evidence, traditional ideas about justification or confirmation of scientific theories are indeed threatened by the character of actual practice in the sciences. Rather than stopping to examine how the hopes of traditional "defensive" epistemology concepts may fare (cf. my 2000) I will outline a different view concerning the demands and norms pertaining to measurement that are operative in scientific practice. The scientific enterprise, conceived as modeling and theorizing subject to the demand of empirical grounding, is a far cry from its traditional philosophical characterization.

4.2 What Counts as Measurement, and What Is Measured?

Undoubtedly theories are tested by confrontation of the empirical implications or numerical simulations of their models with data derived from measurement outcomes.

But for this confrontation to occur, it must first be a settled matter what counts as relevant measurement procedures for physical quantities represented in those models. What counts as the relevant experiment, what counts as measurement, is all that a God-given fact?

On the contrary, the classification of a physical procedure *as measurement* of a parameter in such a model or simulation is itself provided by at least a core of the theory itself. I will support this point by exploring several examples in physics, and then attempt to tease out its consequences for epistemological issues concerning scientific practice. I will argue that

> whether or not a procedure *is a measurement*, and if so, *what it measures*, are questions that have, in general, answers only relative to a theory.

But the fear of skepticism that we see lurking in the insistence on receptivity toward "pure" experience, in e.g. Kosso's insistence on theory-neutrality, can be disarmed, because

> those answers, provided by theory, are part of what allows a theory to meet the stringent requirement of *empirical grounding* (if it can!)

For that to become evident does suppose that we are able to set aside certain traditional foundationalist impulses that have tended to infest popular conceptions of the possibility of confirmation, evidence, and evidential support.

4.3 Examination of Measurement Criteria in Action

When discussing a currently accepted theory and its models for certain phenomena, we are in a position where we can take as already settled and given what the quantities that characterize those phenomena are, what the relations among those quantities that constrain the models are, and what are the physical procedures that count as measurements to determine the values of those quantities. Within this context, the extent to which measurement and theory are entangled will remain hidden. The following examples, drawn from episodes in which the theories were still developing, and the exact identification of the targeted phenomenon was still in question, will bring that entanglement to light.

4.3.1 Galileo Measures the Force of the Vacuum

In his *Dialogue Concerning Two New Sciences* Galileo presented the design of an apparatus to measure the force of the vacuum. Given Galileo's hypothesis concerning the vacuum, this does measure the magnitude of that force, though from a later point of view it is measuring a parameter absent from Galileo's theory, namely atmospheric pressure.

The prevailing opinion concerning the vacuum in Galileo's time was that in nature there is a "horror vacui," that a true vacuum is impossible. Galileo saw some evidence for this view, but reinterpreted that evidence as equally supporting the weaker thesis that indeed, there is an aversion of nature for the vacuum, but it is not an absolute — rather there is a force, the force of the vacuum, that tends to eliminate it by drawing the borders together, and this force has a definite but limited magnitude.

(a) His initial evidence for attractive force of the vacuum:

> If you take two highly polished and smooth plates of marble, metal, or glass and place them face to face, one will slide over the other with the greatest ease, showing conclusively that there is nothing of a viscous nature between them. But when you attempt to separate them and keep them at a constant distance apart, you find the plates exhibit such a repugnance to separation that the upper one will carry the lower one with it and keep it lifted indefinitely, even when the latter is big and heavy (Galilei 1914, 59).

Clearly this adhesion can be brought to an end, though not without difficulty. If indeed the adhesion is due to an attractive force, then the magnitude of that force should be measurable. So Galileo takes the bull by the horns and designs a *measuring instrument*. Presupposing his theory of the force of the vacuum, he presents a procedure for measuring, that is, determining the value of, that force under suitable conditions.

(b) Galileo's design (Galilei 1914, 62, figure 4):

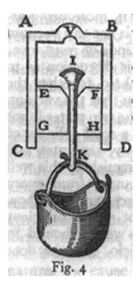

Fig. 4

> The air having been allowed to escape and the iron wire having been drawn back so that it fits snugly against the conical depression in the wood, invert the vessel, bringing it mouth downwards, and hang on the hook K a vessel which can be filled with sand or any heavy

material in quantity sufficient to finally separate the upper surface of the stopper, EF, from the lower surface of the water to which it was attached only by the resistance of the vacuum. Next weigh the stopper and wire together with the attached vessel and its contents; we shall then have the force of the vacuum. (Galilei 1914, 62)

The snug fit of EIF against AVB duplicates the arrangement of the two smooth marble plates. But now this arrangement has been turned into a measuring instrument, with the force measured by the amount of weight it can support, so that a quantitative comparison is made possible.

In retrospect, we do not see things in the same way!

Torricelli's reasoning and more importantly, not much later that century, Pascal's barometer and his experiment on the Puy de Dome, establishes the reality of atmospheric pressure. From that point on, Galileo's instrument has a new theoretical classification: it is still a measuring instrument, but what it measures is a quite different parameter: the force the atmosphere exerts on the surface marked GH in his diagram.

In this case, the instrument is on both sides recognized as a measuring apparatus. But relative to the two different theories, what it measures are two different physical quantities.

4.3.2 Atwood's Machine: Credentialing Newton's Conception

Atwood's machine is often used in class demonstrations and laboratory exercises in the teaching of physics, but its historical role is of much greater interest. This contraption was devised by the Rev. George Atwood, who presented in his book *A Treatise on the Rectilinear Motion and Rotation of Bodies, with a Description of Original Experiments Relative to the Subject* (1784). Some of that history is touched on by Kuhn:

> Consider, for a somewhat more extended example, the problem that engaged much of the best eighteenth-century scientific thought, that of deriving testable numerical predictions from Newton's three Laws of motion and from his principle of universal gravitation. When Newton's theory was first enunciated late in the seventeenth century, only his Third Law (equality of action and reaction) could be directly investigated by experiment, and the relevant experiments applied only to very special cases. The first direct and unequivocal demonstrations of the Second Law awaited the development of the Atwood machine, a subtly conceived piece of laboratory apparatus that was not invented until almost a century after the appearance of the Principia. (Kuhn 1961, 168–169)

Indeed, Atwood's machine was designed to provide measurement results that could test, and confirm, Newton's second law (see below).

But the procedure implemented with this apparatus was interpreted variously also as (a) measuring mass ratios, (b) measuring the force of gravity. Each of these interpretations classify the procedure as measuring *something*, presupposing some parts of Newton's theory, but differ in what they presuppose, and do not classify it the same way.

Let's begin with a description of the apparatus, following Atwood's (op. cit. 299–300) but abstracting from the inevitable falling short of the ideal.

The Machine consists of two boxes, which can be filled with matter, connected by an string over a pulley. The ideal case, modeled most easily, has an inextensible massless string, and the pulley is massless, with zero friction retarding the motion, which occurs in a vacuum.

Result: In the case of certain matter placed in the boxes, the machine is in neutral equilibrium regardless of the position of the boxes; in all other cases, both boxes experience uniform acceleration, with the same magnitude but opposite in direction.

Below is Atwood's Figure 78, depicting his machine.

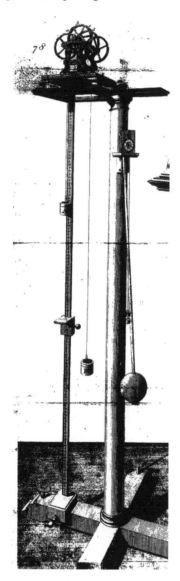

How does this situation look from within Newton's theory? The two objects have masses M and m, say, and are subject to the gravitational force g. If the objects remain at rest, it must be because M = m. If not, the uniform acceleration is due to the force of gravity.

Newton's second law implies that the acceleration equals $g[(M-m)/(M+m)]$. So if the masses are known, and the acceleration measured, then the gravitational acceleration g is determined. That is, presupposing the theoretical classification in terms of mass and force, and assuming the second law, this is an apparatus that measures the force of gravity.

Conversely, if g is known (measured earlier, in a different way, via the acceleration of a freely falling body, also assuming the 2nd law), then measuring the acceleration suffices to determine the mass ratio M/m.

Finally, if both g and the masses are known, and assuming the 3rd law that action = reaction, (tested earlier in a different way by colliding pendulums) then the result *tests* the 2nd law itself.[1]

This explanation of the various arguments that can be constructed around the experimental results helps to understand how various writers, in different historical circumstances, could "read" them in different ways.

4.3.2.1   Interpretation 1: Refuting the Cartesian Objection

It is to be appreciated that Cartesian physics did not die with Descartes, and that Newton's theory too had to struggle for survival, for almost a century. The Cartesian critique of Newtonian physics was that by introducing mass and force, which are not definable in terms of spatial and temporal extension (they are, we say now, dynamic rather than kinematic quantities) Newton had brought back the medievals' occult qualities. For, according to the Cartesians, only quantities of extension are measurable.

The Newtonian response was, in effect, that admittedly what is measured *directly* in any set-up is lengths and durations, but that they could show nevertheless how to measure mass and force.

To be precise, on the assumption that the apparatus is located in a uniform gravitational field, without needing to know the strength of that field. The rationale of this response was thoroughly re-investigated in the nineteenth and early twentieth century by Mach, Duhem, and Poincaré (Mach 1960 Ch. II, section V-1–3; Poincaré 1905/1952, 97–105). Some conclusions can indeed be drawn without presupposing Newton's theory: as Mach points out (*The Science of Mechanics*,

---

[1] Cf. Hanson (1958, 100–102). If g is the acceleration due to gravity, the weight of body m with mass m is mg. The unbalanced force on this body is the difference between the weight and the upward pull F, which is equal and opposite to the upward pull on M. But the unbalanced force on a body equals its mass times its acceleration — which is equal but opposite for the two bodies. So we can solve the equations to yield $(M-m)/(M+m) = a/g$. Both a and g can be determined by clock and ruler measurements, in principle. Given the result, an easy calculation leads to the mass ratio M/m.

Ch. II, section I.16) Atwood's machine shows, and allows to measure directly, the constant acceleration postulated in Galileo's law of falling bodies. But the measurement of Newton's dynamic parameters on a body is an operation that counts as such a measurement only relative to Newtonian theory. To say that the operation measures mass, for example, is to presuppose the applicability of Newton's second and/or third law. So for example the Atwood machine, or measurements by contracting springs, presuppose that the set-up as a whole is a Newtonian system, and the values of the masses are calculated from the observations of kinematic quantities via Newton's laws.

That is a very significant point for us, today. It could not have satisfied the Cartesian, but for us, noting that Newton's theory is hereby satisfying the requirements that are actually in force in scientific practice, it shows us that the Cartesian epistemic constraints are not embodied in scientific methodology.

4.3.2.2 Interpretation 2: Measuring the Postulated Universal Force of Gravity

Newton followed up on his principles of mechanics with a great and audacious postulate: the law of universal gravitation. This postulates that between any two bodies there is an attractive force dependent solely on their masses and the distance between them. The principles of mechanics do not include such an "existence postulate," but they do allow for the design of various procedures that count (relative to themselves) as measurements of that force. Atwood's machine provides one of them, as we saw, predating the more famous experiment by Henry Cavendish in 1798. Cavendish used a torsion balance with lead balls whose inertia (in relation to the torsion constant) he could tell by timing the beam's oscillation. Their faint attraction to other balls placed alongside the beam was detectable by the deflection. But this episode too, is a matter of "reading" the results in a certain way: Cavendish had actually set out to measure the Earth's density, but that involved the effect of gravitation.

4.3.2.3 Interpretation 3: Atwood's Response to the Continental Critics

Why was Atwood intent on finding experimental cash value for Newton's second law? In fact he was responding to the century long disputes about the concept of force and the associated law, that the force is in effect measured by change in velocity:

> Many experiments, however, have been produced, as tending to disprove the Newtonian measure of the quantities of motion communicated to bodies, and to establish another measure instead of it, viz. the square of the velocity and quantity of matter; and it immediately belongs to the present subject, to examine whether the conclusions which have been drawn from these experiments arise from any inconsistency between the Newtonian measures of force and matter of fact, or whether these conclusions are not ill founded, and should be attributed to a partial examination of the subject: but some considerations concerning the principles of retarded motions should premised. (Atwood 1784, 30)

Thus, relying on what he could take to be independent measures of mass, with objects of masses 48 and 50 g attached, and assuming what was known of gravitational acceleration by other means, 980 cm/s^2, Atwood carefully verified that the objects accelerated at the predicted rate. In other words, in Atwood's own hands, the procedure using his apparatus was not simply an acceleration measurement, nor a measurement of the force of gravity, nor of the mass ratio, but rather of the quantitative three-term (force – mass – acceleration) relationship. And this procedure counts as a measurement of that relationship, not by itself and not relative to any theory at all, but relative to Newton's theory, for the measurement of the mass ratio and gravitational force themselves involved Newtonian theoretical calculations.

At this point a skeptic might respond that all we see here is a check on coherence or consistency. That is so, but this point is misleading if left thus blankly stated. For there is no danger here of a self-fulfilling prophecy, for the coherence in question is not just between the theoretical principles, but between them and the empirical data. Even that is an understatement: what is demonstrated here is that the theory is sufficiently advanced to provide for the possibility of determining the values of the theoretically introduced physical quantities.

4.3.3   Michelson and Morley Measure the Relative Speed of Earth and Aether

Today the Michelson-Morley 1887 interferometer experiments take a central place in expositions of Einstein's 1905 theory of relativity, and if their historical role is presented at all, that is mainly to celebrate Einstein's insight.

Einstein's crucial conceptual breakthrough is unquestionable. But the history also displays quite clearly the fact that the questions *what qualifies Michelson and Morley's procedure as a measurement?* and *if so qualified, what was measured?* are answered by theoretical classifications of what goes on in that procedure. To be distinguished are (a) Michelson and Morley's own view of what their procedure achieved, (b) how the result was accommodated in Ritz's emission theory of light, (c) how it could be understood within Lorentz's theory of material contraction in motion, and finally, (d) how it is re-described in Einstein's Special Theory of Relativity.

In fact, today it is normally just discussed to narrate how it was that Einstein came to recognize the following pervasive structural feature of nature:

> the speed of light is a universal constant: light speed is the same in all frames of reference, independent of the direction of travel, the source, or the motion of the source relative to the receiver

This feature of the universe is instantiated any time you use a flashlight or turn on your car headlights etc. But it is not obvious to you under those everyday conditions! Can we now say

> That same pattern, which is instantiated in a confused and disturbed fashion everywhere in nature, is displayed and exemplified in the phenomena created in the Michelson-Morley experiment, and the significance of the experiment for us is precisely that it displays this pattern saliently and clearly

or is that to take as transparent fact what is actually the result's appearance in our assimilated theoretical context? To discuss this, we need to have a look at how the phenomenon displayed in the Michelson and Morley interferometer experiments of 1887 was seen by Michelson and Morley themselves, by Walter Ritz, by Lorentz or Fitzgerald, and finally by Einstein. From the differences between them we'll have to conclude that *it was not the experiment taken in and by itself* that brought the constancy of light into the open.[2] We might put it this way: a measurement was made, and a result obtained, but *what was measured?* And this question is answered *sub specie a theory*. It is just when we look at the displayed phenomenon through theory-colored glasses that we see it as exemplifying a universal pattern in nature, bringing to light some aspect of the "real" structure of our universe.

4.3.3.1   Michelson and Morley's Target: Fresnel's Hypothesis for Light Aberration

Michelson and Morley's 1887 article distinguishes the basic ether/wave theory, call it T, from its augmentation T* by Fresnel's hypotheses to overcome a difficulty with respect to light aberration. The argument they present is, in effect:

> Given T, their apparatus measures the relative velocity of earth and ether, and the measurement outcome determines its value to be 0 (to within limit of accuracy), while the conjunction of T* with that assertion and outcome is inconsistent.

What, precisely, was their target? The problem they were addressing harked back to Bradley's finding in the eighteenth century that, due to the movement of the earth, stars will appear slightly displaced. There had been two explanations of this appearance, provided by the emission theory of light as fast traveling particles, and by the "undulatory" (waves in the ether) theory.

As Michelson and Morley note, the emission theory of light had offered a ready explanation: because of the large distance, we can regard the rays of light coming from the star, and reaching the moving earth at successive times, as parallel to each other, and the Earth as moving at right angles to them.

An analogy that fits well with the emission theory of light is that of walking forward in rain that is falling vertically. From the point of view of a rain drop, you are moving toward it as it falls; equivalently, from your point of view the raindrop is moving toward you with the same speed. (Classical relativity!) So from your point of view it is moving along a path inclined toward your position. Similarly then with the rays of light: the telescope through which you see the star is pointed in *that* direction, the inclined line. If you assume that your telescope is pointing straight up, in the direction of the true source of the light ray, you will be miscalculating.

The rain drop analogy points to the model of light as a stream of particles. But the ether/wave ("undulatory") theory of light can easily adapt this explanation, just

---

[2] I'll draw here on the discussion in Grünbaum (1963, 388–393, 395; 1960), on Shankland et al. 1955, and on Martinez 2004.

as it had shown itself in accord with such other emission-explained facts about light as its straight-line propagation. But another empirical finding had raised a serious problem. In the above explanation, the effect depends solely on the relative speed of approach between light and observer (the tangent of the angle is proportional to the ratio of the two absolute velocities). However, the speed of light was known to be different in different media. That presented a puzzle, as Michelson and Morley record in their opening paragraph:

> [I]t failed to account for the fact proved by experiment that the aberration was unchanged when observations were made with a telescope filled with water. For if the tangent of the angle of aberration is the ratio of the velocity of the earth to the velocity of light, then, since the latter velocity in water is three-fourths its velocity in a vacuum, the aberration observed with a water telescope should be four-thirds of its true value (Michelson and Morley 1887, 335).

Fresnel had accordingly proposed a modification of the undulatory theory by adding two hypotheses: *first*, the ether is supposed to be at rest except in the interior of transparent media, in which *secondly*, it is supposed to move with a velocity *less* than the velocity of the medium in the ratio $(n^2-1)/(n^2)$ where n is the index of refraction.

The second hypothesis Michelson and Morley accept as fully established by Fizeau's famous experiment on the speed on light in different media, as well as some of their own work. So they devised their experiment to test the first hypothesis. What this required was the construction of an instrument that would measure differences in the speed of light along paths at right angles to each other, in a set-up rigidly attached the moving opaque body on which we live: the Earth.

Both the schematic form of the experiment and its null outcome — as well as its reading through Einstein's Special Theory of Relativity — are well enough known that it suffices here to state the modest conclusion they reached:

> It appears, from all that precedes, reasonably certain that if there be any relative motion between the earth and the Luminiferous ether, it must be small; quite small enough entirely to refute Fresnel's explanation of aberration. (Michelson and Morley 1887, 341)

So this procedure was first of all presented as measuring the relative speed of earth and ether, assuming only the basic ether/wave theory of light. The null result, which was to prove so important historically in a very different context, was here presented solely as in disagreement with Fresnel's additional hypothesis.

This illustrates quite well how the questions of whether the experimental apparatus is a measuring instrument, and if so, what it is measuring, are answered relative to a theory. To put it conversely: the ether/wave theory of light had the theoretical resources to design a procedure to determined the value of its theoretical quantity *relative speed of earth and ether.*

In what followed, historically, it was looked at through differently theory-colored glasses.

4.3.3.2   The Result as Seen by Ritz

The Michelson-Morley experiment was repeated at different times, with the earth in different states of motion, approximating different moving inertial frames. In each case, light was terrestrial and moved, within the terrestrial frame, in different directions but along paths of the same length.

Notice that no relation between the motion of light in different frames comes into play. In each case, the result is that the speed is constant in all directions in the current frame. So presented, there are clearly two limitations to this experiment.

First, in that experiment, the light is emitted from a source at rest in the inertial frame of the experiment: terrestrial light. If the speed of light is indeed constant it must be the same regardless of the motion of the source. That was precisely what was contested by the young Walter Ritz, when he offered his emission theory of light in 1908, in an attempt to accommodate the new experimental evidence in a classical framework, without involving a postulated ether.

The Michelson-Morley experiment cannot rule on this since in the light source was attached to the interferometer and so, according to the emission hypothesis, the light's speed would be the same in all directions, in the reference frame of the apparatus. Indeed, Ritz was able to show that a number of optical experiments, all of which had spelled trouble for the ether/wave theory, could be accommodated on his hypothesis (cf. Martinez 2004).

Relevant evidence against Ritz' emission theory of light appeared only with the 1913 astronomical observations of binary stars conducted by Dutch astronomer Willem de Sitter. More conclusively, in 1924 the modified Michelson-Morley experiment was finally performed with light from extraterrestrial sources. Rudolf Tomaschek in Heidelberg used starlight, while Dayton C. Miller in Cleveland used sunlight.[3] Contrary to the expectations of the Ritz theory, they obtained the same results as had been found by using terrestrial light from a source at rest in the frame.

That does not affect the main methodological point: seen from within Ritz' theory, the quantities measured were frame-dependent velocities, for light no different, in that respect, from material projectiles. All theories, and hence theoretical classifications of phenomena, whether natural or created in experiment, are vulnerable to refutation by future evidence. Being so general a point, that does not single out, or dismiss, any particular example.

4.3.3.3 Lorenz Sees It Still Differently

As noted, a second limitation of the Michelson-Morley set-up is that though the light was made to travel in different directions, the paths were of the same length. Lorentz, whose views were contrary to both Einstein's and Ritz's, could see the Michelson-Morley phenomenon as displaying and revealing a pattern quite different from what Einstein took it to be, namely a pattern of material contraction in the direction of motion. In that calculation, the equality of the two paths in the experimental set-up played a role. This theory, still today well known and certainly better known than Ritz, had greater acceptance as a rival to Einstein, and there are long standing discussions of how it could accommodate the results of the troubling

---

[3] For discussion of this exciting experimental episode, see Suppe (1993, 191–193).

optical experiments. Certainly the Michelson-Morley experiment was accommodated, and could be viewed as measuring well-defined classical quantities.

Again, complete clarity had to wait for decades. Finally, in 1932 Kennedy and Thorndike set up an experiment with paths of different lengths, which ruled out Lorentz' calculation.

So how must we regard the Michelson-Morley experiment? After the series of experiments that Michelson and Morley carried out, many theoretical models would later bite the dust, though at the time they provided alternative feasible ways of seeing the result. There were observable phenomena on which all participants could agree. They all agreed on the clock times, the lengths of the arms, the presence or absence of interference fringes and so on. But the content of this common agreement was not sufficient to entail that the occurrence exemplified the constancy and source-independence of the speed of light. Today we are in a different position. *Now it is correct to say*: the experimentally created phenomenon exemplifies the constancy of light, independent of direction or source. But as we say this, we are seeing it through our own theory-colored glasses, those of the theory that survived many trials.

The sustained, continued feasibility of seeing the phenomena through the glasses of that single theory, namely Einstein's, that is the empirical achievement which changed the very form of modern physics.

4.3.4 Quantum Mechanics: What Counts as a Measurement at All?

In each of the examples so far the procedure in question was taken, on all hands, as a measurement and the physical apparatus as a measurement instrument. The question that had only theory-relative answers was about just what it was that was being measured, which physical quantity that the measurement was a measurement of. With the creation of quantum mechanics we arrive at a more significant rupture in the conception of measurement itself.[4]

Heisenberg's uncertainty relations imply a statistical relation between the outcomes of concurrently conducted position and momentum measurements:

> Given two identically prepared ensembles of quantum systems, if A-measurements are performed on one ensemble and B-measurements on the other, then the standard deviations calculated from those two sets of measurement results, will satisfy the relation that their product is less than or equal to a certain constant.

On the face of it, any such statistical relation is compatible with the idea that position and momentum have precise values at all times.

---

[4] Examining this episode I will again draw on an early account by Adolf Grünbaum (1957, 713–715) who was in close touch with the pioneering foundational work of Henry Margenau.

Bohr denied insistently that the Heisenberg uncertainty principle is merely a principle of limited measurability. But the initial arguments by Heisenberg and himself — semi-classical thought experiments — seemed to base this denial on some merely operational incompatibility of what would classically have counted as measurement procedures yielding sharp, simultaneous values of position and momentum. That was challenged sharply, and not only by detractors of the theory, by means of designs for operationally feasible position-and-momentum measurements.

4.3.4.1 Time-of-Flight Measurement

First of all there are the "time of flight" measurements. We should emphasize that this technique makes perfect sense in quantum physics.[5] The technique has been subject to rigorous theoretical analysis (e.g. Feynman 1965, 96–98), and is of common experimental and practical use (cf. e.g. Wcirnar et al. 2000). Thus in time-of-flight mass spectrometry, ions are accelerated by an electrical field to the same kinetic energy with the velocity of the ion depending on the mass-to-charge ratio. The time-of-flight is used to measure their velocity, from which the mass-to-charge ratio can be determined. Such apparatus is commercially available to identify material samples.

As was repeatedly pointed out, when this is used together with a record of the emission and reception of the particles, we can retrospectively assign values for velocity and position at e.g., the time of reception.[6] The use of this technique to design experiments involving a putative measurement of simultaneous sharp position and momentum values appears to be both persistent and recurrent in the literature. Quite recently Freeman Dyson introduced it again, describing it as novel (Dyson 2004), though in fact it was exhaustively analyzed already in the sixties (see e.g. Park and Margenau 1968, 239 and ff).

4.3.4.2 Niels Bohr's Reaction

So operational incompatibility is not at issue. Bohr's next reaction was to point out that the crucial term here is "retrospectively." Those retrospective assignments have no value for predictions, so there is not going to be coming from them any predictions that would confound quantum mechanical predictions:

> Indeed, the position of an individual at two given moments can be measured with any desired degree of accuracy; but if, from such measurements, we would calculate the velocity of the individual in the ordinary way, it must be clearly realized that we are dealing with

---

[5] This is an example discussed by Heisenberg himself (1930, 20).

[6] For comparison, here is another procedure, discussed by Margenau, in which the operations themselves are as nearly simultaneous as we please: a gamma ray microscope is used to obtain a definite position number from an electron and simultaneously, by using waves of suitable greater length as well, a definite momentum number. (Margenau 1950, 376–377, 1958; discussed in Grünbaum, loc. cit.)

an abstraction, from which no unambiguous information concerning the previous or future behavior of the individual can be obtained. (Bohr 1963, 66)

It is true that, indeed, the retrospective judgment does not match any possible quantum mechanical state for the particle.[7] Therefore, within the theory, there can be no prediction based on those putative measurement outcomes. Moreover, Bohr is asserting that the spread in outcomes of subsequent measurements shows that no rule of any sort could improve on this predictive failure.

But however that may be, Bohr's statement is misleading. For it can be plausibly understood as asserting that the procedure in question is *indeed a measurement* of simultaneous position and momentum values, with the qualification just that the outcomes do not have any *practical* value.

A more foundational inquiry leads to a much stronger conclusion: *the procedure does not count as a measurement at all*. It must be emphasized here that, in asserting this, we presuppose that it is theory that decides not only on what is measured, if a measurement is made, but on what counts as a measurement in the first place. And it is the criterion for the latter judgment that is first given true rigor and precision in the foundations of quantum mechanics.

4.3.4.3 First Criterion for Counting as Measurement

The time of flight procedure offered a good example for this analysis, and is analyzed at length, for this purpose, in articles by Margenau (1958) and by Park and Margenau (1968).[8]

The direct measurements in this procedure are all of positions. But a calculation is presented, drawing on these direct measurement results, to yield a value for velocity or momentum. Should this procedure — call it P — be accepted as a true, complex, measurement of momentum? There is one *minimal* theoretical criterion — a *coherence* criterion — that is quite straightforward:

- the theory already provides a theoretical probability distribution for outcomes of momentum measurements given any quantum mechanical state;
- the procedure P in question also admits a quantum mechanical theoretical description that implies a probability distribution for its outcomes, given any quantum mechanical state
- The criterion for P being a measurement of momentum is that these two theoretically calculated probability distributions should coincide for all states

---

[7] This point is not trading on the fact that position and momentum are continuous parameters and therefore have no eigenvalues. For this point holds for discretized versions of these observables (or any pair of discrete conjugate observables) and appropriately coarse discretizations.

[8] While Margenau and Park's analyses are illuminating, I do not agree to the conclusion they advocate, which presupposes that every physical operation which can be designed to yield numbers in some systematic fashion defines a physical quantity, independent of the theory.

This is a coherence condition, it is required on the basis of consistency. If this criterion were not satisfied for a given procedure P, and yet P were counted as a measurement procedure for values of momentum, then the theory would yield inconsistent predictions. Momentum is only an example for this general point.

So here already, with this minimal necessary condition (not yet to be taken as sufficient!) we can see that the question, whether or not a procedure counts as a measurement at all, requests a theoretical answer: the question can only be answered completely relative to a theory.

What about the putative time of flight measurement of momentum then? To begin, at least ideally, the time of flight technique does satisfy this criterion for a measurement of momentum, *for a particular case.*[9] With a particle prepared in a definite position state at time $t = 0$ (definite in the sense that it is localized within a small though finite region — a state with compact support) and a later measurement showing its position then, we find a value for its momentum at time 0. So in this situation we see a sequence of direct position measurements, plus a calculation of a value for momentum for the time of the first position measurement. And, for this state preparation, the predicted probability distribution of outcomes of this procedure is the same as the Born conditional probability for outcomes of momentum measurements on systems in that state.

However, the criterion is stronger: the final words "for all states" are crucial.[10] *We cannot conclude* that momentum can be equated with a function of positions over time, on the basis that the measurement outcome predictions for the two will be the same in a particular sort of case. Specifically, there is no warrant for concluding that the system is in a state similarly "localized" with respect to momentum. The only conclusion that is legitimate is that if the time of flight "measurement" of momentum is performed in a "large enough" collective of systems prepared in that same state, then the distribution of outcomes will be the same as in another such collective subject to regular momentum measurements.

4.3.4.4   Second Criterion for Counting as Measurement

What can we think about that putative simultaneous position *cum* momentum measurement? In fact, just because position and momentum are incompatible

---

[9] In the following sense: if at time $t=0$ the particle has a state represented by a wave function with compact support $(-s, +s)$ then the initial Born probability for outcomes of momentum measurements equals the Born probability of measurements of (mass . position at t)/t in the limit for $t \to \infty$. See Park and Margenau (1968, 240–242) for the calculation.

[10] This point is crucial also for other, similar puzzles that have been offered for the understanding of measurement in quantum theory. Specifically, the correlations in an entangled state of several particles — as in the Einstein-Podolski-Rosen example — have been called upon to design putative measurements yielding simultaneous values for conjugate observables (e.g. Park and Margenau 1968, 245). These designs are disqualified provided we insist that the measurement must be made by means of a procedure whose validity does not depend on the initial state of the measured object; see van Fraassen (1974, 301–303; 1991, 220–221).

observables in quantum mechanics, that theory implies that in general *there can be no functional relationship* between outcomes of any series of position measurements and outcomes of momentum measurements. So we have to distinguish: in the particular case of a freely moving particle, the time of flight procedure is legitimate: it will, according to the theory, present no data that would conflict with the predictions for direct measurement of momentum. But it is not true that this procedure qualifies as a momentum measurement procedure!

Why can we not just conclude that we have a measurement here, with a restricted domain of application? The answer is precisely that if we conclude that, and keep in mind that we have a simultaneous position measurement, then we will also have a measurement of such defined quantities as position + momentum. There are no observables of that sort in the theoretical framework. So then we would have putative measurements which are not measurements of any observables; hence as far as the theory is concerned, not measurements of anything at all, hence not measurements, period.

There is thus also a stronger requirement, besides the above minimal coherence condition. For a procedure to be a measurement, relative to the theory, *there must be a quantity that it measures*. A simple way to make the point is this: for a procedure to qualify as a simultaneous joint measurement of quantities A and B, the theory would (according to the criterion displayed above) have to imply that the probabilities of its outcomes match the joint probabilities assigned to A and B. But if A and B do not commute, the theory affords no joint probabilities for their measurement outcomes. Hence the criterion cannot be satisfied, no matter what that procedure is.

Or again: in the case of elementary quantum mechanics, all physical quantities are represented by Hermitean operators. (I'll make the point for this case, though it can be made also for extensions to other classes of operators, as long as there are non-trivial constraints on the theoretical representation of physical quantities.) If a procedure qualifies as a simultaneous measurement of A and B, then there needs to be such an operator representing the quantity measured. But then any linear function of that quantity, such as A + B, will also be represented by such an operator. As von Neumann already saw, if the operators representing A and B are non-commuting then there will be no such representing operator for A + B. So there cannot be a procedure that can count as a simultaneous measurement of such pairs of quantities.

4.3.4.5 The Criterion Applied to Uses of Entangled States

There is another putative procedure for simultaneous measurement of non-commuting observables, in addition to the "time of flight" argument. Made famous by the Einstein-Podolski-Rosen paradox, it is possible for two systems to form a total system in an entangled state of this sort:

the system composed of particles X and Y is in a pure state that is a superposition of the correlated states $|a(i)\rangle \otimes |b(i)\rangle$, for $i = 1, 2, \ldots$, which is also at the same time a superposition of the correlated states $|a'(i)\rangle \otimes |b'(i)\rangle$

and this is possible though

the values a(i) are values of observable A while the values a'(i) are values of observable A' which does not commute with A, and similarly for values b(i), b'(i) of non-commuting observables B and B'.

When all this is the case, the following holds:

Suppose A is measured on the first particle and value a(k) is found. Then the probability of finding value b(k), if B is measured on the second particle, equals 1.

Suppose B' is measured on the second particle and value b'(k) is found. Then the probability of finding value a'(k), if A' is measured on the first particle, equals 1.

In view of this one could propose the following procedure: measure A on the first particle and B' on the second — if values a(k) and b'(m) are found, declare outcome <a(k), a'(m)> of a joint measurement of A and A' on the first particle.

Just like with the time of flight example, we can cite empirical justification for the claim that this procedure is reliable, for the theory predicts a very stable distribution for the actually found outcome pairs a(k), b'(m) for any given prepared joint state of this sort, and hence also for the "inferred" a(k), a'(k) outcome pairs arrived at by direct measurement plus inference.[11]

But from the point of view of the theory, that complex procedure of measurement plus "inference" is not a measurement procedure at all, for there just is no observable that is being measured at all.[12] First of all, the quantities of the theory are those which appear as parameters or variables within models provided by the theory for the representation of phenomena. Then secondly, whether or not a given procedure counts as a measurement procedure (and whether or not the physical apparatus in use counts as a measurement apparatus) depends on whether there is a quantity of the theory for which this procedure, as modeled within the theory, meets the above criteria.

In general then: not only what a procedure measures, if it is a measurement procedure, but whether it is a measurement in the first place, is a question whose answer is in general determined by theory, not solely by operational or empirical characteristics.

4.4   Empirical Grounding

To the extent that they presume or presuppose independence between theory and evidence, traditional ideas about justification or confirmation of scientific theories are threatened by the conclusions reached here. I will not stop to examine how such

---

[11] It would be no use to cavil at the inclusion of a "paper and pencil operation" in arriving at the outcome value — that is almost a universal characteristic of procedures recognized as measurements. Just think of how Eratosthenes measured the size of the earth, for example.

[12] Park and Margenau (1968) leave open the possibility of saying that there is an observable that is being measured, just not one represented in the theoretical models. But once again the criterion requires that a procedure offered as performing measurements must not be one that just happens to apply properly only to a restricted form of initial states that have very special configurations. In fact, Park and Margenau include a proof (concerning what they name "A-type measurements") that this criterion will be violated for any imagined joint measurement of observables represented by non-commuting observables.

traditional concepts may fare. Instead I will outline a different view concerning the demands and norms pertaining to measurement that are operative in scientific practice. This is a departure from the view that the scientist is engaged in *confirming* theoretical hypotheses, or in justifying belief. In that respect it resembles Frederick Suppe's (1993) view that the scientist is engaged in *credentialing* rather than *confirming* hypotheses, though not in its relation to social constructivism.

That different view that I advocate was clearly, if briefly, spelled out by Hermann Weyl. In slogan form, the demand upon theories is that they be *empirically grounded* (my term) which involves both theoretical and empirical tasks.[13] Weyl's view has not seen much discussion in the literature. The main presentation and pursuit of his view came in Clark Glymour's exploration of what counts as *relevant evidence* (Glymour 1975, 1980). The crafting of a relationship between theory and phenomena is an interplay of theory, modeling, and experiment during which both the identification of parameters and the physical operations suitable for measuring them are determined. I have explored this in a different way elsewhere (van Fraassen 2009), but the above case studies provide instances in which this normative constraint on science is evident, and we can clearly see there how the norm of empirical grounding connects with the present conclusions concerning what counts as measurement, or counts as measurement of what.

4.4.1 Tension Between Logical Strength and Relevant Evidence

Let us begin with an epistemological point that may sound quite paradoxical at first blush:

- logically speaking a weak theory cannot be less likely to be true (or empirically adequate) than any of its stronger extensions,
- but when a theory is still weak, e.g., when it is first proposed, there can in generally be very little or even no evidence relevant to its support.

The reason is that, if there is to be relevant evidence at all, it must be possible to design experiments whose outcomes can furnish evidence. To design such an experiment, one has to draw on the implications of the theory, and a weak theory does not imply very much.

Specifically, when first introduced, a model or theory may involve theoretically postulated physical quantities for which there is as yet no measurement procedure available. This possibility is well illustrated by the advent of the atomic theory in the early nineteenth century. The masses of the atoms or molecules, or their mass ratios,

---

[13] Pages 121–122 of his *Philosophy of mathematics and natural science* (NY: Atheneum 1963; first published in German as *Philosophie der Mathematik und Naturwissenschaft* in 1927) While Weyl does not mention any, there are clear connections to Schlick's demand for "unique coordination" which had been further explored by Reichenbach (1920/1965, Ch. IV; see specifically p. 43).

played a significant part in the models offered for chemical processes, but could not be determined from the measurement data. During that century the theory was developed, various hypotheses were added beginning with Avogadro's, and slowly it became possible to connect theoretical quantities to measurable ones. Such development, simultaneously strengthening the theory and introducing new measurement procedures, is not adventitious or optional: it is a fundamental demand on the empirical sciences.[14]

4.4.2   What Is Empirical Grounding?

There are three parts to the criteria imposed by this normative demand. Two of them were emphasized by Weyl and the third by Glymour. They are:

- *Determinability*: any theoretically significant parameter must be such that there are conditions under which its value can be determined on the basis of measurement.
- *Concordance*, which has two aspects:
  - *Theory-relativity*: this determination can, may, and generally must be made *on the basis of the theoretically posited connections*
  - *Uniqueness*: the quantities must be "uniquely coordinated," there needs to be concordance in the values thus determined by different means.
- *Refutability*, which is also relative to the theory itself:
  - there must be an alternative possible outcome for the same measurements that would have refuted the hypothesis *on the basis of the same theoretically posited connections*.

What we have seen amply illustrated in the above case studies is the necessity, indeed inevitability, of the clause "*on the basis of the same theoretically posited connections*" that appears twice in the above components of the demand for empirical grounding. Determination of the value of a physical quantity, represented in a model of certain phenomena, must be by measurements performed on those phenomena, but with the outcomes related to the model by calculations within the theory itself. The point is brought to light by showing the alternatives in the *meaning* of measurement outcomes relative to different theories. The further point, that there is a theoretical question about whether a given procedure counts as a measurement at all, relates closely to the question of which quantities define the models that the

---

[14] This point has often appeared in the scientific and philosophical literature as demands to "operationalize" theoretical concepts, sometimes in polemics against rival theoretical approaches to a common domain — e.g., between advocates of the atomic theory and those advocating energetics, or between behaviorist and cognitive psychology. Such demands fell into disrepute among philosophers because they typically included the presumption that perfectly theory-neutral evidence could be had, or even that theoretical concepts could be reduced to operational ones. But at heart, and however imperfectly, those demands reflect norms operative in scientific practice.

theory provides. The determinability of the values of those quantities, in principle, subject to the above conditions, is a central norm governing scientific activity, and plays a central role in the sense in which scientific inquiry is empirical inquiry.

References

Atwood, G. (1784). A treatise on the rectilinear motion and rotation of bodies, with a description of original experiments relative to the subject. University of Cambridge. Online at http://gdc.gale.com/products/eighteenth-century-collections-online/. Accessed 14 Jan 2013.

Bohr, N. (1963). *The philosophical writings of Niels Bohr: Vol. 1. Atomic theory and the description of nature.* Woodbridge: Ox Bow Press.

Chalmers, A. (2003). The theory-dependence of the use of instruments in science. *Philosophy of Science, 70,* 493–509.

Dyson, F. (2004). Thought experiments in honor of John Archibald Wheeler. In J. D. Barrow et al. (Eds.), *Science and ultimate reality: Quantum theory, cosmology, and complexity* (pp. 72–89). Cambridge: Cambridge University Press.

Earman, J. (Ed.). (1983). *Testing scientific theories* (Minnesota studies in the philosophy of science, Vol. X). Minneapolis: University of Minnesota Press.

Feynman, R. P. (1965). *Quantum mechanics and path integrals.* New York: McGraw-Hill.

Galilei, G. (1914). *Dialogue concerning two new sciences.* New York: Macmillan.

Glymour, C. (1975). Relevant evidence. *The Journal of Philosophy, 72,* 403–426.

Glymour, C. (1980). *Theory and evidence.* Princeton: Princeton University Press.

Grünbaum, A. (1957). Complementarity in quantum physics and its philosophical generalization. *The Journal of Philosophy, 54,* 713–727.

Grünbaum, A. (1960). Logical and philosophical foundations of the special theory of relativity. In A. Danto & S. Morgenbesser (Eds.), *Philosophy of science* (pp. 399–434). New York: Meridian Books.

Grünbaum, A. (1963). *Philosophical problems of space and time.* New York: Knopf.

Hanson, N. R. (1958). *Patterns of discovery.* Cambridge: Cambridge University Press.

Heisenberg, W. (1930). *The physical principles of the quantum theory.* Chicago: The University of Chicago Press.

Kosso, P. (1989). Science and objectivity. *Journal of Philosophy, 86,* 245–257.

Kuhn, T. S. (1961). The function of measurement in modern physical science. *Isis, 52,* 161–193.

Mach, E. (1960). *The science of mechanics* (T. J. McCormack, Trans.). LaSalle: Open Court.

Margenau, H. (1950). *The nature of physical reality.* New York: McGraw-Hill.

Margenau, H. (1958). Philosophical problems concerning the meaning of measurement in physics. *Philosophy of Science, 25,* 23–33.

Martinez, A. A. (2004). Ritz, Einstein, and the emission hypothesis. *Physics in Perspective, 6,* 4–28.

Michelson, A. A., & Morley, E. W. (1887). On the relative motion of the earth and the luminiferous ether. *American Journal of Science, 3rd ser. 34,* 333–345.

Park, J., & Margenau, H. (1968). Simultaneous measurability in quantum theory. *International Journal of Theoretical Physics, 1,* 211–283.

Poincaré, H. (1905/1952). *Science and hypothesis.* New York: Dover.

Reichenbach, H. (1920/1965). *The theory of relativity and a priori knowledge* (Maria Reichenbach, Trans.). Berkeley: University of California Press.

Shankland, R. S., McCuskey, S. W., Leone, F. C., & Kuerti, G. (1955). New analysis of the interferometer observations of Dayton C. Miller. *Reviews of Modern Physics, 27,* 167–178.

Suppe, F. (1993). Credentialling scientific claims. *Perspectives on Science, 1,* 153–203.

van Fraassen, B. C. (1974). The Einstein-Podolsky-Rosen Paradox. *Synthese, 29,* 291–309.

van Fraassen, B. C. (1983a). Glymour on evidence and explanation. In J. Earman (Ed.), *Testing scientific theories* (Minnesota studies in the philosophy of science, Vol. X, pp. 165–176). Minneapolis: University of Minnesota Press.

van Fraassen, B. C. (1983b). Theory comparison and relevant evidence. In J. Earman (Ed.), *Testing scientific theories* (Minnesota studies in the philosophy of science, Vol. X, pp. 27–42). Minneapolis: University of Minnesota Press.

van Fraassen, B. C. (1991). *Quantum mechanics: An empiricist view*. Oxford: Oxford University Press.

van Fraassen, B. C. (2000). The false hopes of traditional epistemology. *Philosophy and Phenomenological Research, 60,* 253–280.

van Fraassen, B. C. (2009). The Perils of Perrin, at the hands of philosophers. *Philosophical Studies, 143,* 5–24.

Wcirnar, R., Romberg, R., Frigo, S., Kasshike, B., & Feulner, P. (2000). Time-of-flight techniques for the investigation of kinetic energy distributions of ions and neutrals desorbed by core excitations. *Surface Science, 451,* 124–129.

Weyl, H. (1927/1963). *Philosophy of mathematics and natural science*. New York: Atheneum.

Chapter 5
On Representing Evidence

Maria Carla Galavotti

Abstract This contribution addresses a number of issues related to the representation, use and appraisal of evidence, with a special focus on the health sciences and law. It is argued that evidence is a trans-disciplinary notion whose distinctive trait is its capacity to provide a link between some body of information and some hypothesis such information supports or negates. As such, evidence is strictly associated with relevance, and like relevance it is intrinsically context-dependent. An analysis of evidence has to address a number of issues, including the epistemic context of reference, the general or particular nature of the hypothesis under scrutiny, the predictive or explanatory character of the inference in which evidence is involved, and the stage at which a given body of evidence is being used within a complex inferential process. Moreover, an awareness of the context in which evidence is appraised recommends that all assumptions underlying the representation of evidence be rigorously spelled out and justified case by case, and the ultimate aims of evidence be clearly specified.

Keywords Evidence • Scientific inference • Explanation • Prediction • Manipulation

5.1 Foreword

The notion of evidence has recently become the object of increasing attention from researchers in various disciplines, and has generated an extensive literature devoted to the clarification of its nature and inferential uses.

By contrast, evidence has only recently become a subject field for philosophers of science. This is due to a long-standing consensus on the clear-cut distinction between a context of discovery and a context of justification, dating back to the birth

M.C. Galavotti (✉)
Department of Philosophy, University of Bologna, Zamboni Street, 38, 40137 Bologna, Italy
e-mail: mariacarla.galavotti@unibo.it

W.J. Gonzalez (ed.), *Bas van Fraassen's Approach to Representation and Models in Science*, Synthese Library 368, DOI 10.1007/978-94-007-7838-2_5, © Springer Science+Business Media Dordrecht 2014

of philosophy of science in connection with the Vienna and Berlin Circles. Such distinction is described by Hans Reichenbach as: "the well-known difference between the thinker's way of finding this theorem and his way of presenting it before a public [...] I shall introduce the terms *context of discovery* and *context of justification* to mark the distinction. Then we have to say that epistemology is only occupied in constructing the context of justification" (Reichenbach 1938, 1966[6], 6–7). The idea behind it is to keep the sociological and psychological aspects of theory formation separate from the precision and rigour characterizing the final formulation of theories. While the sociological and psychological components of the process leading to the statement of a theory belong to the context of discovery, *rational reconstruction*, namely the process aiming "to have thinking replaced by justifiable operations" (*ibid.*, 7) is the object of the context of justification. Logical empiricists identify the goal of philosophy of science with the "rational reconstruction" of scientific knowledge, namely the clarification of the logical structure of science, through the analysis of its language and methods. By identifying justification as the proper field of application of philosophy of science they intended to leave discovery out of its remit; the context of discovery was then discarded from philosophy of science and left to sociology, psychology and history.

The distinction between context of discovery and context of justification goes hand in hand with the tenet that the theoretical side of science should be kept separate from its observational and experimental components. The final, abstract formulation of theories should be analyzed apart from the process behind it, including the complex methodology for the collection and organization of empirical findings. In other words, the "plane of observation," including all that comes from observation and experimentation, is taken as given, and is not to be analyzed, like all that belongs to the context of discovery and not to that of justification.

The view of theories upheld by logical empiricists, together with the distinction between the context of discovery and the context of justification, has gradually been superseded by a more flexible viewpoint according to which theory and observation are intertwined rather than separate, as are the contexts of discovery and justification. Such a change in perspective was triggered by the pioneering work of Patrick Suppes who, starting with his article "Models of Data," which appeared in 1962, and in a long series of subsequent writings culminating in the monumental book *Representation and Invariance of Scientific Structures* (2002),[1] opened philosophy of science to the study of the context of discovery as an integral part of scientific knowledge. Suppes's perspective marks an about-turn with respect to the received view developed by logical empiricists, which he contrasts with a pragmatist standpoint that regards theory and observation as intertwined rather than separate, establishes a continuity between the context of discovery and the context of justification, and takes scientific theories as principles of inference useful for making predictions and choosing between alternative courses of action.

A crucial aspect of Suppes's approach is the acknowledgment that "empirical structures," namely the models organizing and describing empirical data, are objects

---

[1] See also the collection of papers in Suppes (1993).

of investigation no less important than logical structures. This opens the door to a whole array of issues concerning observation, experimentation, measurement, and statistical methodology for collecting data and assessing their bearing on scientific hypotheses. Aware of the importance of these components of scientific method, Suppes insists that philosophy of science is concerned as much with formal logic and set theory as with probability and statistical inference, and labels his own perspective "probabilistic empiricism," to stress the crucial role played within epistemology by probability.

Suppes's viewpoint is deeply pluralistic, in the conviction that the tendency to look for univocal accounts and solutions typical of logical empiricism should be abandoned in favour of a multi-faceted and context-sensitive view of scientific knowledge. In this spirit, Suppes calls attention to the complexity of data delivered by observation and experimentation. In his words: "the 'data' represent an abstraction from the complex practical activity of producing them. Steps of abstraction can be identified, but at no one point is there a clear and distinct reason to exclaim, 'Here are the data!'" (Suppes 1988, 30). Depending on the desired level of abstraction different pieces of information will then count as "data," and what qualifies as "relevant" will inevitably depend on a cluster of context-dependent elements. In what follows it will be argued that Suppes' emphasis on the complex nature of data and the need to take into account the context in which one operates should be extended to the broader notion of evidence.

Suppes is not alone in heralding a context-sensitive approach to epistemology. In recent years a similar tendency has been embraced by a number of authors including Bas van Fraassen — to whose work the present volume is devoted. Both Suppes and van Fraassen paid great attention to measurement, as well as to the relationships between models of data and theoretical models. In addition to physics, the main focus of van Fraassen's research, Suppes addressed learning theory and more recently the structure of the brain. By contrast, the present contribution focusses on the health sciences and law, two fields attracting growing attention on the part of those interested in foundational issues.

5.2 Evidence as a Multi-disciplinary Subject

According to the Oxford Dictionary, evidence is "anything that gives reason for believing something; that makes clear or proves something." Evidence can consist of information of various kinds including empirical data coming from observation and experiment, images, oral reports, recordings, and materials of different sorts. All such types of evidence raise serious problems of collection, representation and interpretation. The awareness of the role played by evidence in the process of establishing and assessing hypotheses in all branches of science, and also in everyday life, is the focus of lively debate among researchers active in several fields.

The jurist William Twining, a leading protagonist in that debate, maintains that "all disciplines that have important empirical elements are connected to a shared family of problems about evidence and inference. Apart from its theoretical interest (as a contribution to human understanding) evidence is of great practical importance in many spheres of practical decision-making and risk management. In particular, multi-disciplinary study of evidence focuses attention on such questions as: (i) What features of evidence are common across disciplines and what features are special? (ii) What concepts, methods and insights developed in one discipline are transferable to others? (iii) What concepts are not transferable? Why? (iv) Can we develop general concepts, methods and insights that apply to evidence in all or nearly all contexts?" (Twining 2003, 97). Such questions are the core of extensive research done in recent years fostering the conviction that evidence is a "multi-disciplinary subject in its own right" (*ibid.*, 99), and one can speak of a *science of evidence*.[2] This conviction goes hand in hand with the awareness that both the production and interpretation of evidence raise peculiar problems within different contexts. While in some scientific fields, such as physics, one relies on "hard" data, often collected according to protocols approved by the scientific community, in others, like medicine and law, what counts as evidence "cannot be restricted to 'hard' scientific data" (*ibid.*, 96).

In an attempt to identify the trans-disciplinary nature of evidence, Twining claims that "at its core, evidence as a multi-disciplinary subject is about inferential reasoning" (*ibid.*, 97). In other words, the distinctive trait of evidence is identified with its capacity to provide a relation between some body of information and some hypothesis that is supported or negated by it. As such, evidence is strictly associated with the notion of *relevance*.

The analysis of evidence has to take into account a number of issues, including the epistemic context of reference, the general or particular nature of the hypothesis under scrutiny, the predictive or explanatory character of the inference in which evidence is involved, and the stage at which a given body of evidence is being used within a complex inferential process. In the course of an insightful discussion of the use of evidence in the realm of law, Twining maintains that "in considering problems of evidence and inference three distinctions are crucial: the difference between *past-directed* and *future-directed* inquiries; the distinction between *particular* and *general* inquiries; and the distinction between *hypothesis formation* and *hypothesis testing*" (*ibid.*, 103; italics added). Twining's distinctions are crucial, and bear directly on the discussion developed in the following sections.

Also important with regard to evidence is *classification*. This is strongly emphasized by David Schum, a pioneer of the science of evidence, who claims that "being able to classify evidence on inferential grounds has many useful consequences. This allows us to discuss some very general properties of evidence and to meaningfully compare the meaning of evidence in different evidential reasoning tasks and

---

[2] Questions of this kind have been the focus of the interdisciplinary research supported by Leverhulme Foundation "Evidence, inference and enquiry: Towards an integrated science of evidence," carried out between 2004 and 2007 under the guidance of the statistician Philip Dawid. This research project led to the publication of Dawid et al. eds. (2011b).

within a given particular inferential task" (Schum 2011, 13). Schum puts forward a "substance-blind classification of evidence" meant to apply to the analysis of evidence independently of its particular content, and therefore in a trans-disciplinary fashion. Schum distinguishes three major dimensions of evidence: *relevance*, *credibility*, and *inferential force or weight*. The relevance dimension has to do with the bearing of evidence upon the hypothesis that has to be proved or disproved. In that connection, evidence can be *direct* or *indirect*, depending on whether it can be related to the hypothesis by a "defensible argument or chain of reasoning," in which case it is direct, or "it bears upon the strength or weakness of links in a chain of reasoning set up by directly relevant evidence" (*ibid.*, 20), in which case it is indirect. The credibility dimension has to do with how those who evaluate evidence stand in relation to it. In other words, it concerns the question: "can we believe that the event(s) reported in the evidence actually occurred?" (*ibid.*, 21). Schum regards this as the most complex aspect of evidence because "we must ask different credibility-related questions for different kinds of evidence we have" (*ibidem*). A first distinction that matters in connection with this dimension of evidence is between *tangible* and *testimonial* evidence, where the first can be examined directly, while the second is reported by testimonies. These two kinds of evidence obviously raise a number of problems such as authenticity, reliability and accuracy in the case of tangible evidence; competence, veracity and credibility in the case of testimonial evidence, where the credibility of a witness also involves his veracity, objectivity and observational ability. No less complex is the assessment of the inferential force or *weight* of evidence. Part of the problem is that there is no general consensus on how weight should be defined and assessed. A number of different views and methods have been developed by statisticians belonging to different schools, but as Schum remarked "no single view says all there is to be said about the force or weight of evidence" (*ibid.*, 23) because this would require other elements to be considered in addition to statistical measures. In fact "the force or weight of evidence depends on assessments made regarding the other two evidence credentials: relevance and credibility" (*ibidem*). For instance, one would have to consider the strength of the links of a chain of reasoning brought to sustain the relevance of a given body of evidence for a certain hypothesis, or the credibility of its source.

Having said that, it should be added that evidence has a lot to do with statistics. As stated by Leonard Jimmie Savage: "statistics consists in trying to understand data and to obtain more understandable data" (Savage 1977, 4). Statisticians developed a vast array of statistical methods for collecting and organizing evidence (descriptive statistics), for inferring various kinds of conclusions from evidence (inferential statistics), and for testing hypotheses against data. Granted that statisticians prompted powerful and useful tools, their application raises myriad problems. As emphasised by C. G. G. Aitken: "scientific evidence requires considerable care in its interpretation. There are problems concerned with the random variation naturally associated with scientific observations. There are problems concerned with the definition of a suitable reference population against which concepts of rarity or commonality may be assessed. There are problems concerned with the choice of a measure of the value of evidence" (Aitken 1995, 4). Evidence is often employed to specify causal knowledge that goes beyond mere statistical correlations. It is vital to acknowledge that this requires assumptions that should be based on solid grounds and justified case by case.

Also worth noting is the fact that exploiting and accumulating evidence may sometimes involve ethical issues. This is obviously true in the realm of medicine. Experimenting the efficacy of a new treatment, for example, requires careful evaluation of potential risks, which often proves problematic. In order to test the safety and efficacy of a new treatment researchers carry out experiments, usually applying randomization techniques. The adoption of randomization in medicine is itself the object of ongoing debate, (see for instance Worrall 2006) but even apart from that the evaluation of the risks faced by individuals who agree to undergo experimental treatments depends on myriad factors that need to be considered with great care. This holds both for the risks to which the individuals who accept to undergo experiments are exposed, and for the risks to which the population at large is exposed once a drug is made available or a surgical treatment enters medical practice. In order to answer questions like: "What are the risks of a potential new treatment for liver cancer? Are the risks outweighed by the potential clinical benefits? What dose of the treatment is best?" (Rid and Wendler 2010, 151), one has to assess the possibility to generalize the results of experiments. Obviously, this procedure involves not only technical, but also ethical and practical issues that can only be appraised within a given context.[3]

5.3   Evidence in the Health Sciences[4]

The health sciences cover a diversified range of sub-disciplines including epidemiology, clinical medicine, pathology, anatomy, and so on, all of which pursue different purposes. Epidemiology is involved with devising practices to avoid or reduce the risk of spreading diseases, while clinical medicine aims at diagnosis and therapy, and pathological anatomy aims at reaching knowledge of the human body that can explain the insurgence of diseases. To such tasks there corresponds a nonuniform involvement with prediction, manipulation, and explanation, which is usually taken in its causal meaning as knowledge of the mechanisms responsible for diseases. The accomplishment of all of these conceptual operations obviously needs to be supported by evidence. The health sciences make extensive use of statistical relationships, but often evidence concerning single individuals is also required, for instance to adjust some therapy to a given patient. The distinction between information regarding whole populations and information regarding individuals is therefore of the utmost importance in this setting.

---

[3] See for instance a recent issue of the journal *Law, Probability, and Risk*, 9 (2010), n. 3–4, entirely devoted to "Risk and probability in bioethics."

[4] This section benefits from joint work with Raffaella Campaner.

The foundations of the health sciences are the object of growing concern for philosophers of science. Among those who have made substantial contributions to the debate on the topic Federica Russo and Jon Williamson argue in the course of a discussion of the nature of causality in medicine that "the health sciences make causal claims on the basis of evidence *both* of physical mechanisms, and of probabilistic dependencies" (Russo and Williamson 2007, 157). So far so good, but they go on to claim that "there are not two varieties of cause but two types of evidence" (*ibid.*, 166). The two kinds of evidence that matter in medicine according to Russo and Williamson are *probabilistic* and *mechanistic* (see also Russo and Williamson 2011). While it is undeniable that both mechanistic and probabilistic evidence play a fundamental role in the establishment and assessment of causal hypotheses in the health sciences, this classification cannot be taken as exhaustive because there is at least one more kind of evidence that matters, namely *manipulative evidence*. Moreover, probabilistic and mechanistic evidence should be seen as complementary rather than opposed. According to a vast literature dating back to the 1970s and constantly growing ever since, mechanisms can be conceived in probabilistic terms, so that probabilistic evidence expressed by means of correlations can and often does suggest mechanisms. As Salmon clearly stated, the identification of mechanisms requires more than statistical correlations, but these represent the first step in the search for mechanisms. Evidence of correlations is apt to direct interventions that may prove useful to find out about mechanisms, which suggests that evidence can be of a manipulative kind.

The crucial role played by evidence provided by manipulations has been pointed out by various authors including Paul Thagard, who in the course of a discussion of the hypothesis that Helicobacter pylori causes ulcers emphasizes the relevance of evidence from manipulative interventions, namely evidence that "eradicating bacteria cures ulcers" (Thagard 1998, 132) for the acceptance of that hypothesis (for more on this see Campaner 2011, 12).

Evidence in the health sciences is also discussed by Jeremy Howick, Paul Glasziou and Jeffrey Aronson, who speak of "evidence hierarchies" and distinguish among *direct evidence* "from studies (randomized and non-randomized) that a probabilistic association between intervention and outcome is causal and not spurious," *mechanistic evidence* "for the alleged causal process that connects the intervention and the outcome," and *parallel evidence* "that supports the causal hypothesis suggested in a study, with related studies that have similar results" (Howick et al. 2009, 186). The authors also mention *evidence for mechanisms* to refer to evidence provided by statistical correlations that hints at the existence of some mechanism.

The same point is emphasized by epidemiologist Paolo Vineis, who calls attention to the fact that preventive measures in epidemiology are sometimes achieved "in the absence of any clue as to the biological causes or mechanisms of action" (Vineis and Ghisleni 2004, 203).

To sum up, both *manipulative* and *mechanistic* evidence are essential to medical research and practice, where they are deeply intertwined. Probabilistic evidence qualifies as transversal rather than opposite with respect to other kinds of evidence, and the same holds for direct and indirect (or parallel) evidence.

The distinction between manipulative and mechanistic evidence is paralleled by the distinction between two similar concepts of causality coexisting in a number of recent accounts, including those put forward by James Woodward, Stuart Glennan, Peter Machamer, Lindley Darden and Carl Craver (see Woodward 2003, 2004; Glennan 2002, 2010; Machamer et al. 2000). The author of the present pages also endorsed a pluralistic view of causality apt to accommodate both of these notions and suggested they could be combined within the "perspectival" approach of Huw Price, which relates causality to the agent's perspective, holding that to call A a cause of B is to regard A as a potential means for achieving the end B (see Price 1991, 2007). Price's epistemic approach can be taken to provide a broad philosophical framework that "in order to become a flesh and blood theory of causality [...] has to be substantiated by more specific accounts" (Galavotti 2001, 8. See also Galavotti 2008). The nature of such accounts will inevitably depend on the context, more particularly on the aims of the enquiry being conducted and on the kind of evidence available. The perspectival viewpoint is fully compatible with the idea that whenever mechanistic evidence is available on that ground mechanistic hypotheses and models can be devised.

While playing a fundamental role, causal analysis in medicine is characterized by a high degree of complexity. A case study that gives an idea of such a complexity is provided by deep brain stimulation (DBS), a therapeutic technique employed to suppress tremors in patients with advanced Parkinson's disease.[5] DBS consists in a surgical operation which inserts components for electric stimulation, targeted mainly at the subthalamic nucleus or the globus pallidus. High-frequency stimulation produced by the electrodes causes a functional block of the anatomic structure, and, by blocking electrical signals from targeted areas in the brain, reduces the hyperactivity responsible for Parkinson's disease symptoms. Remarkably positive long-term effects and advantages are largely documented, whereas side-effects and complications are rare and disturbances are transient. Difficulties are mainly due to the complexity of the phenomenon under examination, and are amplified by the reactions of patients: a wide range of strictly personal aspects, such as the conformation of the skull, age, possible reactions to drugs, psychological attitude, and others, are regarded as responsible for a marked variability in responses. Such difficulties notwithstanding, DBS is being increasingly employed for Parkinson's and a number of other diseases such as dystonia, Tourette syndrome, depression and obsessive compulsive disorder. While DBS is effective in many cases, details are largely unknown about *why* it is so and what the *exact processes* are. In other words, researchers have not managed to decipher *how* DBS brings about its effects. Thus DBS exemplifies a case in which therapy not only precedes but contributes to the discovery of mechanistic details. While "the precise mechanisms of action for DBS remain uncertain, [...] mapping the effects of this causal intervention is likely to help us unravel the fundamental mechanisms of human brain function"

---

[5] This example, which I owe to Raffaella Campaner, is discussed in more detail in Campaner and Galavotti (2007, 2012).

(Kringelbach et al. 2007, 623), and to clarify fundamental issues such as the functional anatomy of selected brain circuits and the relationships between activity in those circuits and behaviour. It is worthwhile stressing that such a technique is leading to progress in elucidating not only the neural mechanisms directly underlying the effects of DBS, but also the fundamental brain functions affected in the targeted brain disorders. In the absence of mechanistic knowledge, causation can be conceived of as manipulation, both for practical and heuristic purposes. So Kringelbach et al. (2007) explicitly speak of "the causal and interventional nature" of DBS, and discuss various different hypotheses that have been put forward to account for the underlying mechanism.

Knowledge of mechanisms is what researchers aim at, because once mechanisms are known disease can be explained on that basis. This can be done either in terms of a mechanism at work or in terms of a mechanism's impairment. Moreover, mechanistic knowledge allows for making prediction and planning manipulation. In the case of manipulation, however, a distinction should be made between interventions to be performed at the *population level* like those planned by the epidemiologist, and interventions on *single individuals* like therapies (pharmaceutical, surgical, etc.). These two cases call for *different kinds of evidence*, since the first makes use of statistical data referred to populations, while the second also requires information on individual patients.

Causal analysis can also be conducted at different levels, so that one can have *general* or *type causality* (referred to populations), and *singular* or *token causality* (referred to individuals). This distinction has a long tradition within the literature on causation due to statisticians. Irving John Good, for instance, grounded his theory of probabilistic causality on this distinction, while Philip Dawid has repeatedly called attention to it more recently (see Good 1961–1962; Dawid 2000, 2007). The distinction lies at the basis of Salmon's two levels of explanation, namely the *statistical-relevance model* according to which events are explained by locating them in a network of statistical relations holding between the properties relevant to their occurrence, and *mechanical* explanation in terms of *processes* and *interactions*, which is meant to explain single events by exhibiting the (probabilistic) mechanisms responsible for their occurrence. Salmon regards the shift from type-level analysis to token-level analysis as relatively unproblematic. However, while this may be true of physics, the major field of application of Salmon's theory, it surely does not hold for other disciplines, including psychology, medicine, and the social sciences.[6] As a matter of fact, the shift from types to tokens is highly problematic in the health sciences, and requires great care.

Evidence available in medicine often does not allow a complete description of the mechanisms at work, and use is made of only partially specified mechanisms. This is emphasized by a number of authors including Peter Machamer, Lindley Darden and Carl Craver who speak of *mechanism schemas* and *sketches*, and

---

[6] This is admitted by Salmon himself in (2002). For more on Salmon's theory of explanation and causality see Salmon (1984, 1998). See also Galavotti (2010) where Salmon's theory is discussed in the framework of the broader debate on explanation.

Donald Gillies who refers to *plausible mechanisms* (see Machamer et al. 2000; Gillies 2011). The search for mechanisms in medicine is usually articulated into a multi-level analysis requiring both mechanical and manipulative evidence, referring to populations as well as individuals. This is exemplified by the DBS case, where use is made of *general (statistical) evidence* as well as *particular information*, and both *past-directed* and *future-directed* inquiries are conducted. In fact, a multi-layered analysis is performed involving mechanisms at upper and lower levels (motions disorders, chemical deficiencies, electrical transmission of signals), and the effects of manipulation across such levels are investigated.

As already observed, evidence can serve various purposes in the health sciences. In epidemiology evidence is accumulated for the sake of *prediction* and *policy interventions*. Epidemiological analysis is conducted at some level of generality and evidence is expressed by means of statistical correlations because what matters are average values rather than data concerning the individual members of a population. Statistical correlations to be employed for prediction and interventions have to be *robust*, namely they have to be invariant, or stable across a broad range of varying conditions and circumstances. The degree of robustness required from such correlations will depend on the use to which the predictions obtained on their basis are to be put, as well as on the kind of interventions that are being planned, their cost, risk, urgency, and so on. By contrast, *interventions* in clinical medicine are made on *single patients*, and in addition to statistical correlations evidence regarding individuals is needed. When the available evidence suggests that some fully or partially known mechanism is at work, the physician makes a diagnosis and plans a therapy. At that stage, in most cases additional evidence, often manipulative in kind, is required to adjust the therapy, or to decide upon further steps to be taken. Different yet again is the case of autopsy, where what is sought is an explanation of why somebody died requiring both general and individual information, and causal analysis is typically *ex-post*.

It is worth calling attention to the assumptions that are (often tacitly) made whenever evidence, especially statistical evidence, is used for prediction, planning interventions, and establishing causal connections. One extensively adopted assumption is *invariance across different regimes*, typically *observational* and *interventional* — or experimental (with or without randomisation). As recommended by Philip Dawid, a statistician who devoted great attention to the analysis of evidence, assuming invariance across regimes requires great care. The issue intertwines with the distinction between *general (type)* and *singular (token)* causal analysis, because the task of type analysis, as described by Dawid, is to use past data to make choices about future interventions, and "this requires that we understand very clearly the real-world meaning of terms such 'observational regime' and 'interventional regime', since there are many possible varieties of such regimes" (Dawid 2007, 529). This can only be accomplished with reference to the context in which one operates. As Dawid put it: "appropriate specification of context, relevant to the specific purposes at hand, is vital to render causal questions and answers meaningful" (Dawid 2000, 422). Dawid's advice to spell out all assumptions that are made and to justify them case by case invokes once again the centrality of context.

5.4   Evidence in Law

The nature, role and evaluation of evidence in the realm of law is the focus of extensive debate. Evidence is generally employed in law to support *analysis ex-post*, and has to do with the appraisal of *particular hypotheses*. In Twining's words: "adjudication of issues of fact in contested trials is typically past-directed, particular, and hypothesis testing" (Twining 2011, 88). In addition, "disputed trials are typically concerned with inquiries into particular past events in which the hypotheses are defined in advance by law — what lawyers call 'materiality'. Moreover, records of cases are artificially constructed units extracted from more complex and diffuse contexts. For example, a criminal trial may be just one event in a long-drawn out feud or other conflicts. These elements — particularity, pastness, materiality, and individuation of cases — differentiate this kind of legal material from many other inquiries in which reasoning from evidence is involved" (*ibid.*, 88–89). A further element characterizing evidence in law amounts to the fact that in adjudication a decision has to be taken, and "this pressure for decision has led the law to develop important ideas about presumptions, burden of proof and standards of proof as aids to decision" (*ibidem*).

The study of evidence in law has benefitted from the proliferation and refinement of techniques for identification by means of fingerprints, DNA evidence, marks on bullets, etc.; the ever-increasing amount of epidemiological and medical data, and the progress of risk analysis. The organization and appraisal of evidence is entrusted to forensic scientists, who make use of it for the sake of identification, for instance to identify the source of a trace left at a murder scene. The method employed to accomplish this task is *comparison*. Typically, evidential material found at the scene of a crime is compared with other evidential material found, say, on a suspect's clothing, or in his car. Statistics provides the means for making such comparisons. As C. G. G. Aitken observed: "statistics has developed as a subject, one of whose main concern is the quantification of the assessments of comparisons. The performance of a new treatment, drug or fertilizer has to be compared with that of an old treatment, drug or fertilizer, for example. Statistics and forensic science are increasingly interacting thanks to the increasing amount of available data (DNA, refractive index of glass fragments, chromatic coordinates measuring colour in fibres, etc.)" (Aitken 1995, 16). The goal of this kind of comparison is to help those who are in charge to make a judgment in a variety of situations ranging from paternity disputes to the judgment of innocence or guilt in case of a criminal offence. To be sure, the final judgment is up to judges and/or jurors, and usually requires a whole array of considerations of a different sort, such as causal knowledge, to mention one. The attribution of responsibility is ruled by different standards in tort and criminal law: in tort law the standard is *preponderance of probability*, while criminal law demands the BARD (*Beyond A Reasonable Doubt*) standard. How to relate the probabilistic representations of evidence obtained by means of statistical methods to a concept like the BARD principle raises delicate problems and fosters endless debate.[7]

---

[7] These and other related issues are addressed in Redmayne (2001).

A major problem lurking behind the application of statistical methods is the identification of an appropriate *reference class*. Ideally, a suitable reference class for base rates should be such that no relevant variables are omitted (to avoid confounding) and that data are carefully collected. This obviously creates a problem that admits of no simple and general solution, and can only be addressed in a context-sensitive fashion.[8]

In the 1970s Dennis Lindley launched the adoption of Bayesian methodology as a tool apt to help decision-making in court. His work started a trend in the literature that has burgeoned ever since. At the core of Lindley's proposal lies the *likelihood ratio* (LR), taken as an optimal measure of the *value of the evidence* with respect to competing hypotheses. The hypotheses considered can be various. For instance, in a paternity dispute they might sound like "the alleged father is the true father of the child" and "the alleged father is not the true father of the child"; and in a murder case one might have the following: "the material found at the crime scene came from a Caucasian" and "the material found at the crime scene came from an Afro-Caribbean".

Such competing hypotheses may also be those of guilt and innocence of a defendant, in which case the LR compares the weight of a given body of evidence under the hypothesis that a suspect has committed a crime and the alternative hypothesis that he did not commit that crime. Some care is needed when probability is applied to this kind of hypotheses. Lindley calls attention to the fact that when probability is applied to the hypothesis of guilt it refers "to the event that the defendant committed the crime with which he has been charged [...] not to the judgment of guilt" (Lindley 1991, 27). The *hypothesis* of guilt should not be conflated with the *judgment* of guilt, which falls within the competence of judges or jurors, who ground it on a complex body of information not reducible to mere quantitative evidence. The same point is stressed by Aitken, who claims that "it is very tempting when assessing evidence to try to determine a value for the probability of guilt of a suspect, or a value for the odds in favour of guilt and perhaps even reach a decision regarding the suspect's guilt. However, this is the role of the jury and/or judge. It is not the role of the forensic scientist or statistical expert witness to give an opinion on this" (Aitken 1995, 4).

Not itself a probability, the LR results from comparing two probabilities, namely the probability of the evidence E given the hypothesis H and the probability of E given the hypothesis G:

$$LR = p(E \mid H) / p(E \mid G)$$

or, to weigh a body of evidence with respect to a given hypothesis and its negation:

$$LR = p(E \mid H) / p(E \mid -H).$$

---

[8] The literature on statistics in law reflects an increasing awareness of the importance of this problem. See for instance Taggart and Blackmon 2008.

The LR relates naturally to the notion of *relevance*, in the sense that a LR of value 1 means the given body of evidence is irrelevant to the hypothesis, whereas a value that differs from 1 suggests that the given body of evidence is relevant. More particularly, a likelihood ratio greater than 1 indicates how much a given body of evidence favours the truth of a certain hypothesis against the alternative under consideration, and conversely if the likelihood ratio is less than 1. A number of authors including Evett, Robertson and Vignaux define as "weak" for adoption in court a likelihood ratio in the range 1–33, "fair" a ratio in the range 33–100, "good" a ratio in the range 100–330, "strong" a ratio in the range 330–1,000, and "very strong" a ratio greater than 1,000 (Robertson and Vignaux 1995, 12. See also Evett 1991).

Although the LR has a meaning of its own, Bayesians recommend its use within the Bayesian framework, where it plays a crucial role in connection with the shift from prior to posterior probabilities. This appears evident if Bayes' rule is expressed in terms of odds:

$$\left[p(H \mid E) / p(-H \mid E) \right] = \left[p(H) / p(-H) \right] \times \left[p(E \mid H) / p(E \mid -H) \right].$$

By considering the shift from prior to posterior probabilities one can evaluate how a given body of evidence is apt to influence the comparison between two hypotheses by favouring one of them against the other. A very high value of the LR can convert a low prior probability into a high posterior probability. Just to give an idea of the effect of the LR on the shift from prior to posterior probability, a LR = 100 would transform a prior of 0.5 into a posterior of 0.99. Supposing that one wanted to apply Bayes's reasoning to the two hypotheses of guilt and innocence of a defendant, given a body of evidence estimated (through the LR) to be 100 times more likely conditional on the guilt than on the innocence hypothesis, to obtain a posterior probability of at least 99 % — that is to say a value apt to satisfy the BARD standard (see Lindley 1975) — one would need a prior probability, namely the probability of guilt before that body of evidence is taken into account, of at least 50 %. Clearly, in case a certain trace or single item E were the only evidence, it could lead to a probability value of 99 % only if combined with a very strong likelihood ratio. As Dawid observed, "when E is the only evidence in the case, before E is admitted the suspect should be treated no differently from any other member of the population, and then a prior probability of guilt of even 1 in 1,000 could be regarded as unreasonably high" (Dawid 2005b). Obviously, fixing the value of priors is a most delicate operation involving several considerations not amenable to quantitative analysis. For this reason, a number of authors recommend the application of the Bayesian method at an advanced stage of the trial.

Representing evidence by means of the LR proves fruitful not only in court, but also in medicine and many other fields. Obviously, the use of the LR is beset with difficulties, and the same holds for Bayes's rule, namely because there is no unique recipe for calculating likelihoods, precisely as there is no univocal way of fixing priors. For these and other reasons a number of authors favour the adoption of the methods of classical statistics, like tests of significance and tests of hypotheses, rather than Bayesian methodology. The use of statistical methods in court is matter of hot debate, and the literature on the topic is constantly growing.[9]

Regrettably, statistics have often been misused in court. A case in point is the widespread argument known as the *prosecutor's fallacy*. An instance of this fallacy, which can take various forms, obtains when a *match probability*, namely the probability that a given piece of evidence such as a trace left at a murder scene is to be ascribed to an individual taken at random from a reference population, is taken as the probability that the defendant is not guilty, and then the conclusion is drawn that the probability of his guilt is $(1 - p)$. Take for instance a match probability $p \, (M \mid -G) = 1/10,000,000$, where $M =$ a trace found at the murder scene, and $-G =$ the defendant is not responsible for it, namely the trace was left by an individual chosen randomly from the reference population. The fallacy obtains by confusing the match probability $p \, (M \mid -G)$ with $p \, (-G \mid M)$, namely the probability that the defendant is not guilty given the piece of evidence found at the murder scene, and then drawing the conclusion that the probability of the defendant being guilty is $1 - 1/10,000,000$. In this way a very high probability of guilt of the defendant is derived from a very low probability, based on the fallacious move known as *transposing the conditional*.[10] The prosecutor's fallacy exemplifies the intricacies that surround the adoption of probabilistic reasoning in court. As Dawid put it, "seemingly straightforward problems of legal reasoning can quickly lead to complexity, controversy and confusion" (Dawid 2005b).[11]

The challenges posed by probabilistic reasoning and the complexity characterizing evidence in most cases can make statistical calculations very laborious and the process leading from evidence to a certain conclusion remain opaque. Moreover, it is often problematic to make probability values obtained by experts as the result of inferences from complex bodies of evidence understood to those who have the responsibility to take decisions based on them, like jurors and judges, but also doctors, epidemiologists, and decision-makers operating in different fields. To deal with such difficulties a number of techniques for the graphical representation of evidence and evidence-based reasoning have been developed. A landmark in the literature on the topic is John Henry Wigmore's *The Science of Judicial Proof as Given by Logic, Psychology, and General Experience, and Illustrated in Judicial Trials*, which appeared in 1913. In this work, that can be traced back to the rationalist tradition dating back to Jeremy Bentham, Wigmore develops the so-called *chart method*, meant as a "rigorous system that enables and requires the lawyer to identify and to appraise possible logical relationships that evidential data may be argued to have to intermediate and ultimate propositions that must be proved in a particular

---

[9] Some of the objections to the use of probability and statistics in court are discussed in Galavotti (2012). For a discussion of Bayesian methods in the law see Fienberg and Finkelstein (1996). An interesting comparison between the Bayesian and frequentist approaches to a DNA identification problem is to be found in Kaye (2008).

[10] For an extensive discussion of the prosecutor's fallacy see Gigerenzer (2002).

[11] Dawid (2005b) examines a few examples of the problems arising in the field, and contains a useful list of bibliographical references. See also Dawid (2002).

case. It requires that the propositions and the relationships claimed to exist among them be articulated and recorded in a systematic manner that makes it easier to criticize and appraise each step in an argument and the argument as a whole" (Anderson and Twining 1991, 329–330). The chart method, subsequently revised and extended by Terence Anderson, William Twining, David Schum and others, starts from a distinction between *factum probandum*, expressed by a proposition to be proved, and *factum probans*, describing the evidence relevant to that proposition, and is meant to represent the inferential relationships between single pieces of circumstantial and testimonial evidence and *probanda*. According to Dawid, a Wigmore chart "focuses on inference towards some ultimate probandum, emphasizes the distinction between occurrence and report of an event, pays particular attention to the many links in a chain of reasoning, and assists qualitative analysis and synthesis" (Dawid 2008, 143). Schum labeled the method "relational structuring" to stress its power to illustrate "the typically catenated, cascaded, or hierarchical nature of arguments" (Schum 1993, 178).

An alternative method for representing the relationships between evidence and hypotheses of interest is given by *Bayesian networks*. These are extensively used by forensic scientists to address complex problems involving mixed or indirect evidence, with the support of appropriate software. Applied to a given problem, like a case of disputed paternity, a Bayesian network can "describe the probabilistic relationships between the variables involved, enter evidence on some of them, and 'propagate' this to obtain revised probabilities for other variables" (Dawid 2008, 137). In general, Bayesian networks are used to represent causal dependencies among variables, under appropriate assumptions.[12] As described by Dawid, both Wigmorean charts and Bayesian networks "organize many disparate items of evidence and their relationships, focus attention on required inputs, and support coherent narrative and argumentation" (*ibid.*, 142). To be sure, neither of these approaches is intended to give "objective" representations of reality, being rather meant to reflect the viewpoint of somebody like the prosecutor, or the defense lawyer.[13] Typically, they are addressed to those in charge of making a judgement as an aid to see both the reasoning that lies behind a certain conclusion and the evidence brought in its favour. Moreover, "by using reach hierarchically structured representations human reasoners can overcome the limitations imposed by their limited-capacity working memory" (Lagnado 2011, 202). Although graphical methods of representation have been developed mostly in connection with legal evidence, attempts to extend their application to a broader range of problems are under study. Major developments in that connection are likely to be achieved in the near future.

---

[12] For an extensive treatment of Bayesian networks and their use in forensic science see Taroni et al. (2006).

[13] This is emphasized in Dawid et al. (2011a), which contains a detailed comparison of Bayesian and Wigmorean networks.

5.5 Concluding Remarks

The topic of evidence is obviously much broader than suggested here. As emphasized in the first section, evidence is gaining increasing attention from researchers and decision-makers operating in fields other than the health sciences and law. The preceding remarks were meant to give an idea of the importance of the topic and the complexity that surrounds it. If a conclusion can be taken from our discussion, it amounts to an acknowledgment of the centrality of context. More particularly, an awareness of the context in which one operates recommends that all assumptions underlying the representation of evidence are rigorously spelled out and justified case by case. Similarly, the aims to which evidence is to be put should be specified. Within the health sciences, this holds especially in connection with explanation, prediction, and manipulation. It is also important to classify the nature of the available data and clarify the nature of the inferential links between evidence and hypotheses.

References

Aitken, C. G. G. (1995). *Statistics and the evaluation of evidence for forensic scientists*. Chichester: Wiley.

Anderson, T., & Twining, W. (1991). *Analysis of evidence*. London: Weidenfeld and Nicholson.

Campaner, R. (2011). Understanding mechanisms in the health sciences. *Theoretical Medicine and Bioethics, 32*, 5–17.

Campaner, R., & Galavotti, M. C. (2007). Plurality in causality. In P. Machamer & G. Wolters (Eds.), *Thinking about causes* (pp. 178–199). Pittsburgh: University of Pittsburgh Press.

Campaner, R., & Galavotti, M. C. (2012). Evidence and the assessment of the causal relations in the health sciences. *European Studies in the Philosophy of Science, 26*, 27–45.

Dawid, P. (2000). Causal inference without counterfactuals. *Journal of the American Statistical Association, 95*, 407–424.

Dawid, P. (2002). Bayes's theorem and weighing evidence by juries. In R. Swinburne (Ed.), *Bayes's theorem, proceedings of the British Academy 113* (pp. 71–90). Oxford: Oxford University Press.

Dawid, P. (2005a). Probability and statistics in court. Appendix online to the second edition of T. Anderson, D. Schum, & W. Twining (Eds.), *Analysis of evidence*. Cambridge: Cambridge University Press. http://tinyurl.com/7q3bd. Accessed 14 Jan 2013.

Dawid, P. (2005b). Probability and statistics in the law. In R. G. Cowell & Z. Ghahramani (Eds.), *Proceedings of the tenth international workshop on artificial intelligence and statistics (AISTATS 2005)*. http://tinyurl.com/br8fl. Accessed 14 Jan 2013.

Dawid, P. (2007). Counterfactuals, hypotheticals and potential responses: A philosophical examination of statistical causality. In F. Russo & J. Williamson (Eds.), *Causality and probability in the sciences* (pp. 503–532). London: College Publications.

Dawid, P. (2008). Statistics and the law. In A. Bell, J. Swenson-Wright, & K. Tybjerg (Eds.), *Evidence* (pp. 119–148). Cambridge: Cambridge University Press.

Dawid, P., Schum, D., & Hepler, A. (2011a). Inference networks: Bayes and Wigmore. In P. Dawid, W. Twining, & M. Vasilaki (Eds.), *Evidence, inference and enquiry. Proceedings of the British Academy 171* (pp. 119–150). Oxford: Oxford University Press.

Dawid, P., Twining, W., & Vasilaki, M. (Eds.). (2011b). *Evidence, inference and enquiry. Proceedings of the British Academy 171*. Oxford: Oxford University Press.

Evett, I. W. (1991). Interpretation: A personal odyssey. In C. G. G. Aitken & D. A. Stoney (Eds.), *The use of statistics in forensic science* (pp. 9–22). New York: Ellis Horwood.

Fienberg, S., & Finkelstein, M. O. (1996). Bayesian statistics and the law. In J. M. Bernardo, J. O. Berger, P. Dawid, & A. F. M. Smith (Eds.), *Bayesian statistics* (pp. 129–146). Oxford: Oxford University Press.

Galavotti, M. C. (2001). Causality, mechanisms and manipulation. In M. C. Galavotti, P. Suppes, & D. Costantini (Eds.), *Stochastic causality* (pp. 1–14). Stanford: CSLI.

Galavotti, M. C. (2008). Causal pluralism and context. In M. C. Galavotti, R. Scazzieri, & P. Suppes (Eds.), *Reasoning, rationality and probability* (pp. 233–252). Stanford: CSLI.

Galavotti, M. C. (2010). Probabilistic causality, observation and experimentation. In W. J. Gonzalez (Ed.), *New methodological perspectives on observation and experimentation in science* (pp. 139–155). A Coruña: Netbiblo.

Galavotti, M. C. (2012). Probability, statistics, and law. In D. Dieks, W. J. Gonzalez, S. Hartmann, M. Stoeltzner, & M. Weber (Eds.), *Probability, laws, and structures* (pp. 401–412). Dordrecht: Springer.

Gigerenzer, G. (2002). *Reckoning with risk: Learning to live with uncertainty*. New York: Simon and Schuster.

Gillies, D. (2011). The Russo-Williamson thesis and the question of whether smoking causes heart disease. In P. M. Illari, F. Russo, & J. Williamson (Eds.), *Causality in science* (pp. 110–125). Oxford: Oxford University Press.

Glennan, S. (2002). Rethinking mechanical explanation. *Philosophy of Science, 69*, S342–S353.

Glennan, S. (2010). Mechanisms, causes and the layered model of the world. *Philosophy and Phenomenological Research, 81*, 362–381.

Good, I. J. (1961–1962). A causal calculus. Part I and II. *British Journal for the Philosophy of Science, 11*, 305–318; *12*, 43–51; Errata and Corrigenda *13*, 88.

Howick, J., Glasziou, P., & Aronson, J. (2009). The evolution of evidence hierarchies: What can Bradford Hill's 'guidelines for Causation' contribute? *Journal of the Royal Society of Medicine, 102*, 186–194.

Kaye, D. (2008). Case comment – *People v. Nelson*: A tale of two statistics. *Law, Probability and Risk, 7*, 249–257.

Kringelbach, M. L., Jenkinson, N., Owen, S., & Tipu, A. (2007). Translational principles of deep brain stimulation. *Nature Review Neuroscience, 8*, 623–635.

Lagnado, D. (2011). Thinking about evidence. In P. Dawid, W. Twining, & M. Vasilaki (Eds.), *Evidence, inference and enquiry. Proceedings of the British Academy 171* (pp. 183–224). Oxford: Oxford University Press.

Lindley, D. V. (1975). Probabilities and the law. In D. Wendt & C. Vlek (Eds.), *Utility, probability, and human decision making* (pp. 223–232). Dordrecht/Boston: Reidel.

Lindley, D. V. (1991). Probability. In C. G. G. Aitken & D. A. Stoney (Eds.), *The use of statistics in forensic science* (pp. 27–50). New York: Ellis Horwood.

Machamer, P., Darden, L., & Craver, C. (2000). Thinking about mechanisms. *Philosophy of Science, 67*, 1–25.

Price, H. (1991). Agency and probabilistic causality. *British Journal for the Philosophy of Science, 42*, 157–176.

Price, H. (2007). Causal perspectivalism. In H. Price & R. Corry (Eds.), *Causation, physics, and the constitution of reality. Russell's republicanism revisited* (pp. 250–292). Oxford: Clarendon Press.

Redmayne, M. (2001). *Expert evidence and criminal justice*. Oxford: Oxford University Press.

Reichenbach, H. (1938). *Experience and prediction*. Chicago/London: Chicago University Press. 6th edition 1966.

Rid, A., & Wendler, D. (2010). Risk-benefit assessment in medical research – Critical review and open questions. *Law Probability and Risk, 9*, 151–177.

Robertson, B., & Vignaux, R. (1995). *Interpreting evidence*. Chichester: Wiley.

Russo, F., & Williamson, J. (2007). Interpreting causality in the health sciences. *International Studies in the Philosophy of Science, 21*, 157–170.

Russo, F., & Williamson, J. (2011). Generic versus single-case causality: The case of autopsy. *European Journal for the Philosophy of Science, 1,* 47–69.

Salmon, W. C. (1984). *Scientific explanation and the causal structure of the world.* Princeton: Princeton University Press.

Salmon, W. C. (1998). *Causality and explanation.* New York/Oxford: Oxford University Press.

Salmon, W. C. (2002). A realistic account of causation. In M. Marsonet (Ed.), *The problem of realism* (pp. 106–134). Aldershot: Ashgate.

Savage, L. J. (1977). The shifting foundations of statistics. In R. Colodny (Ed.), *Logic, laws, and life* (pp. 3–18). Pittsburgh: University of Pittsburgh Press.

Schum, D. A. (1993). Argument structuring and evidence evaluation. In R. Hastie (Ed.), *Inside the juror* (pp. 175–191). Cambridge: Cambridge University Press.

Schum, D. A. (2011). Classifying forms and combinations of evidence: Necessary in a science of evidence. In P. Dawid, W. Twining, & M. Vasilaki (Eds.), *Evidence, inference and enquiry. Proceedings of the British Academy 171* (pp. 11–36). Oxford: Oxford University Press.

Suppes, P. (1962). Models of data. In E. Nagel, P. Suppes, & A. Tarski (Eds.), *Logic, methodology and philosophy of science* (pp. 252–261). Stanford: Stanford University Press.

Suppes, P. (1988). Empirical structures. In E. Scheibe (Ed.), *The role of experience in science* (pp. 23–33). Berlin/New York: Walter de Gruyter.

Suppes, P. (1993). *Models and methods in the philosophy of science: Selected essays.* Dordrecht/Boston: Kluwer.

Suppes, P. (2002). *Representation and invariance of scientific structures.* Stanford: CSLI.

Taggart, A., & Blackmon, W. (2008). Statistical base and background rates: The silent issue not addressed in *Massachusetts v. EPA. Law, Probability and Risk, 7,* 275–304.

Taroni, F., Aitken, C., Garbolino, P., & Biedermann, A. (2006). *Bayesian networks and probabilistic inference in forensic science.* Chichester: Wiley.

Thagard, P. (1998). Ulcers and bacteria I: Discovery and acceptance. *Studies in History and Philosophy of Science. Part C: Studies in History and Philosophy of Biology and Biomedical Sciences, 29,* 107–136.

Twining, W. (2003). Evidence as a multi-disciplinary subject. *Law, Probability and Risk, 2,* 91–107.

Twining, W. (2011). Moving beyond law: Interdisciplinarity and the study of evidence. In P. Dawid, W. Twining, & M. Vasilaki (Eds.), *Evidence, inference and enquiry. Proceedings of the British Academy 171* (pp. 73–118). Oxford: Oxford University Press.

Vineis, P., & Ghisleni, M. (2004). Risks, causality and the precautionary principle. *Topoi, 23,* 203–210.

Wigmore, J. H. (1913). *The science of judicial proof as given by logic, psychology, and general experience, and illustrated in judicial trials.* Boston: Little Brown, and Co., 3rd ed. 1937.

Woodward, J. (2003). *Making things happen: A theory of causal explanation.* New York: Oxford University Press.

Woodward, J. (2004). Counterfactuals and causal explanation. *International Studies in the History and Philosophy of Science, 18,* 41–72.

Worrall, J. (2006). Why randomize? Evidence and ethics in clinical trials. In W. J. Gonzalez & J. Alcolea (Eds.), *Contemporary perspectives in philosophy and methodology of science* (pp. 65–82). A Coruña: Netbiblo.

Part III
Models and Reality

Chapter 6
Scientific Models of Abduction: The Role of Non Classical Logic

Ángel Nepomuceno

Abstract Inspired by van Fraassen's viewpoint, according to which logic is relevant to philosophy of science, in this chapter we aim to lead the way of applying logic to the study of abduction. Certainly van Fraassen seems to reject abduction, but many logicians have considered abduction as a prototype of scientific inference, and, at the last resort, this critical position is more about the so called inference to the best explanation than about abduction itself. We recover the original notion, due to Peirce, as an inferential process that permits to formulate explicative hypotheses. A couple of examples of abduction in scientific practices are given. Then we present the classical model of logical treatment of abduction (the AKM-model), its connection with the AGM-model of belief revision and its limits. In both cases the logical parameter is a classical logic, but a change of such parameter is methodologically justified: semantic tableaux are very illustrative, since some new rules could be necessary to obtain closed branches, that is to say, to obtain the corresponding keys to solve abductive problems. Then a dynamic perspective may be adopted, from which multimodal logics are suitable, so we study Bonano's system with epistemic operators and another one with four modal operators.

Keywords Abduction • Scientific practices • Non classical logic • Belief revision • Multimodal logic

This work has been carried out as part of the research project *Alternative Interpretations of Non-Classical Logics*, reference HUM5844 of the Ministry of Economy, Innovation and Science (Junta de Andalucía), and the project *Consciousness, Logic and Computation*, reference FFI2011-29609-C02-01 of the Ministry of Science and Innovation (Government of Spain).

Á. Nepomuceno (✉)
Department of Philosophy and Logic, University of Seville,
Camilo José Cela Street, w/n, 41018 Seville, Spain
e-mail: nepomuce@us.es

W.J. Gonzalez (ed.), *Bas van Fraassen's Approach to Representation and Models in Science*, Synthese Library 368, DOI 10.1007/978-94-007-7838-2_6,
© Springer Science+Business Media Dordrecht 2014

6.1 Theoretical Framework

van Fraassen somehow sustains a semantic point of view about how philosophy of science could be understood and it has been pointed out that his constructive empiricism is a well developed alternative to scientific realism. Sometimes the border-line between logic and philosophy of science, particularly when the latter is seen as methodology, is hard to detect, though many philosophers of science do not share this idea. On the contrary, van Fraassen considers a certain relevance of logic to philosophy of science and, at the same time, as a consequence of his criticism about scientific realism, he seems to reject abduction, which has been considered by many logicians as a prototype of scientific inference. However, our author has worked productively not only in philosophy of science, but in classical logic and extensions of classical logic, which means an exceptional position to tackle this kind of interdisciplinary subjects.

Deduction and induction have been traditionally considered as the inference forms *par excellence* in scientific research, the former in the context of justification, as a guarantee of greater certainty, and the later as a trend to advance knowledge in the context of discovery. A deduction does not lead to new knowledge, any deductive conclusion has information which is contained in the information of premises, so we do not get new information. On the contrary, in an inductive generalization this is very different. For example, in an inductive generalization we have more information in the conclusion than in the inductive premises. On the other hand, abduction is the typical way of increasing the initial information.

In this chapter we aim to lead the way of applying logic to study abduction, which is inspired in van Fraassen's opinion about the relevance of logic to philosophy of science, which is not inconsistent with his criticism over abduction, since it is restricted to some of its forms. In particular, we shall emphasize the role of non classical logic, in consonance with the fact that sometimes classical entailment relation is not the only one, as van Fraassen himself has detected.

The work is organized as follows. After this introductory section, there is a section to establish, from the original approach of Peirce, the argumentative nature of the abduction process. van Fraassen's possible criticism is not about abduction in general. In the section that follows some cases of abduction in scientific practices are presented, one on paleoanthropology and the other in a field of social sciences and humanities, linguistics. Further, paying attention to a suggestion by van Fraassen, we study the logic of abduction, then the classical model is tackled, which requires to take into account some logical concepts and define different styles of abduction. This is related to the AGM theory of epistemic change, and we outline the method of semantic tableaux as a resource for calculating classical abductive solutions. A later section is devoted to present two modal models, one multimodal and a S4 system, also comparable with the AGM theory, to finish the work with some concluding remarks and a short bibliography.

6.2 Abduction as a Form of Inference

Logic becomes a branch of the theory of argumentation that studies a class of inferences, precisely those that are considered correct (valid) in certain areas of knowledge, such as scientific research or reasoning to make decisions, artificial intelligence, etc. We will use 'inference,' 'argument' and 'reasoning' as though they were synonymous, though strictly speaking their three respective meanings are not completely equivalent.

The result of an inferential process is a sequence of propositions, which begins with an initial set, the premises, a final proposition, the conclusion, and between them other propositions are placed (in some cases, absent), the 'chain of reasons.' The pair (premises/conclusion) represent the argument of inference, so that, in Fregean terms, the inference can be understood as a dyadic function whose arguments are the aforementioned pair.

Deduction is the form of inference in which it is intended that the truth of the conclusion follows from the truth of the premises, that is to say that a situation in which the premises are true and the conclusion is false is unacceptable. When the truth of the conclusion follows from the premises, we say that the inference is valid. The induction, however, presents some premises such that if they are true, then the conclusion is probably true, so there is no question of validity of inductive inference, but more or less force to consider the conclusion true.

On the other hand, abduction is a reasoning by means of which one tries to explain a puzzling observation. It is the explicative reasoning par excellence, which is present in several practices, as common sense, medical diagnosis and scientific reasoning. Then, we could ask if it is a form of induction or another form of reasoning that is present in such practices. But one thing at a time, and let us see the Peircean point of view about that.

Peirce, as we said, adds abduction as another form of inference (Peirce 1991). This notion has been considered different from deduction and induction and it is defined by means of the following rule (Peirce 1998):

The surprising fact, *C*, is observed.
But if *A* were true, then *C* would be a matter of course.
Hence, there is reason to suspect that *A* is true

Naturally, such conclusion from the first two premises does not appear in the same manner as that of the familiar fallacy of affirming the consequent, the conclusion in the fallacy is made as it was a deductive reasoning, but the truth of the conclusion does not follow from the premises. Now, however, it does not proclaim with certainty (the abductive conclusion), but *there is reason for A to be true*, i.e., what is actually concluded is that *A* is plausible.

To compare with the first two, we can see some patterns in the line proposed by the American author (Peirce 1998)

- Deduction: rule + case = result
  All *A* is *B*, this *a* is *A*, so that *a* is *B*

- Induction: case + result = rule
  This *a* is *A*, this *a* is *B*, then all *A* is *B*
- Abduction: rule + result = case
  All *A* is *B*, this *a* is *B*, then it is plausible to consider that this *a* is *A*

Following this terminology, we can say that abductive inference, from an abductive problem, which is given by a rule and a result, intends to find the case, i.e., if the premises, the abductive problem constituents, are true, then the proposed hypothesis must be plausible. This hypothesis is the result of the process of abductive inference (the conclusion of such an inference).

From another point of view, abduction is considered as a kind of induction. Already Peirce himself (Peirce 1998), well versed in the history of traditional logic, underlies his conception of abduction founded on the Aristotelian concept of *apagogé*, to get the "induction universalizing" (*epagogé*), as Beuchot noted (2002). At this point it is interesting that to understand the abduction like induction is even prior to Peirce. In this regard, as shown in Martínez-Freire (2010).

> abduction is authentic induction, i.e., the empirical scientific reasoning searched for (and not found) by Francis Bacon[1]

and, moreover, it is well studied but does not receive this name, in Whewell, a previous author to Peirce, who proposed it as an alternative to the induction of Mill, in a work bearing the suggestive title *Novum Organon Renovatum*. This approach does not detract from the thesis given below, so in the last resort, abduction can be seen as a specific form of inference.

Because of the richness of this Peircean concept, research about abduction is extensive. A logical treatment is given in the so called AKM-model,[2] to which we shall pay more attention below. From another perspective, in Gabbay and Woods (2006) and Woods (2012), the GW-schema is proposed. This schema is based on the concept of *ignorance problem*, according to which if you want to know a proposition *A,* and you do not have enough information to answer this question, or to draw it out by implication (or projection) from your knowledge, then you have an ignorance problem with respect to *A*.

From another point of view, to study abduction, pragmatics considerations should be taken into account, in fact it may be creative, revolutionary and the thing may be to choose the most coherent set of hypotheses and not to restrict explanations to cases in which all the data are to be explained (Thagard and Shelly 1997). An extended eco-cognitive approach to abduction, not logically oriented, is given in Magnani (2009).

All of that shows a strong interdisciplinary character of abduction, as it was to be expected, giving the variety of epistemic scenarios in which abductive problems arise (in the next section two examples corresponding to two different scenarios of knowledge are presented). Nevertheless, we are interested in the logical treatment

---

[1] "La abducción es la inducción auténtica, es decir, el razonamiento científico empírico buscado (y no encontrado) por Francis Bacon," in the original.

[2] Aliseda, Kakas or Kowalski, and Meheus or Magnani (Gabbay and Woods 2006).

of abduction, and the existence of diverse approaches also shows the necessity of going beyond classical logic and analyzing models and representations in a dynamic perspective.

What is Peircean abduction, after all? In Hintikka (1999), abduction, or inference of hypotheses for testing or verification, is labeled as the fundamental problem of contemporary epistemology, and five theses (initially due to Kapitan) summarizes the main features of abduction. Such theses are:

1. *Inferential thesis.* Abduction is, or at least includes, a process or inferential processes
2. *Thesis of purpose.* The purpose of scientific abduction is both (i) to generate new hypotheses, and (ii) to select hypotheses for further examination and testing, hence a central issue in scientific abduction is to provide methods for selecting
3. *Comprehension thesis.* Scientific abduction includes all operations whereby theories are engendered
4. *Autonomy thesis.* Abduction is, or embodies, reasoning that is distinct from, and irreducible to, either deduction or induction

Sometimes the thesis of purpose has been underlined, particularly the methods of selecting hypotheses, and it is considered that abduction is no more than *inference to the best explanation*, IBE to abbreviate (Lipton 1996), so that any logical treatment should account for the best possible explanation, although the selection methods are very difficult to formalize.

Some defenses in favor of scientific realism against instrumentalism are in fact variants of IBE, as it is pointed out by Psillos (1996), then van Fraassen's point of view should be critique of that kind of reasoning. In order to analyze this perspective, Psillos identifies two kinds of IBE: *horizontal* IBE, which involves only hypotheses about unobserved but observable entities, and *vertical* IBE that involves hypotheses about unobservable entities. According to Psillos, van Fraassen does not doubt horizontal IBE and his point of view would be against vertical IBE, which is based on his dichotomy between observables and unobservables (van Fraassen 1980), distinction that plays an epistemic role.

However, such distinction does not avoid van Fraassen's objections to both notions, horizontal and vertical IBE, since van Fraassen gives two arguments against the so understood abduction (van Fraassen 1989): argument from the bad lot, according to which in order to advocate of IBE, to argue that IBE leads to truth a principle of privilege must be assumed, and argument from indifference, which is based on the assumption that the only thing known about the best explanation is that it belongs to the (probably infinite) class of theories that explains the corresponding fact. But Psillos (1996) insists that such arguments fail to undermine vertical IBE.

Then what is controversial is a kind of abduction.

From our perspective IBE does not exhaust the concept of abduction, which is rather the inference to *some* explanation in order to clarify something that was previously presented as surprising fact, and not a simple inference, but a richer process, as the aforementioned thesis establishes. So that the previous discussion is, in the last resort, more about conditions to take an inferential process to be relevant as to formulate an explicative hypothesis. However, the bounded explicative power of a hypothesis abductively obtained does not invalidate the notion itself. On the other hand, though any logical treatment of abduction would be very difficult, it cannot be restricted to IBE, which could be seen as a metatheoretic idealization of the corresponding processes.

Whatever the case may be, Peirce believes that there is a form of inference that leads to the formulation of *plausible* hypotheses and, therefore, it is one of the forms of scientific inference. There are other ones, though we do not deal with other inferences that are certainly important in scientific research, such as statistical inference, or analogical reasoning, for example, since the three proposed types offer the basic needed elements to tackle the most interesting aspects of the methodology of science.

6.3   Abduction in Some Scientific Practices

First we refer a case of modern paleoanthropology.[3] In the development of a work in the *Sierra de Atapuerca* (Burgos, Spain), between 1994 and 1995, a group of researchers discovered a set of 80 fossils of hominids, with some remains of animals and 200 very primitive tools. The facial skeleton of a young individual was particularly interesting. These findings were in a layer (called *Aurora Stratum*) of about 25 cm, at level 6 of the so called *Trinchera Dolina* (in short, TD6), and the investigation determined that the remains are those of the late Pleistocene, so having a history of over 780,000 years.

Everything seems to be normal. We have a series of data that result from the exhaustive search for information in a relatively recent site of great importance in paleoanthropological research. However, the discovery was a surprising fact in connection with the dominant paradigm at that time, a population of such hominids was not expected to exist in Europe in this age. In fact, in Western Europe the known oldest hominids are the ones of the specie *Homo Heidelbergensis* (appeared 500,000 years ago), which, according to current knowledge, was the common ancestor of *Neanderthals* (230,000 years) and *Sapiens* (150,000 years).

The age of TD6 fossils settles a problem about their phylogenetic classification and their location in the tree of evolution, despite a cranial capacity of 1,000 cc and their jaw, with evolutionary connections with European and African forms from Middle and Upper Pleistocene. The face had not the primitive features that are characteristic of *Homo Ergaster* (1.7 million years), nor was it a line derived from *Homo Erectus* (proceeding from the former), neither was it derived from *Homo Heidelbergensis* or *Homo Neanderthalensis,* its successor. But they were virtually the features of *Homo Sapiens*, hence it was definitely a surprising fact.

---

[3] We follow the example given in a personal communication by A. Rivadulla, supplemented with data by Arsuaga (2001) and it has been presented in a methodological course (Master in Evolutive Biology, University of Sevilla). Moreover, the notion of *preduction* (Rivadulla 2007), used in the case of physics, has been slightly modified.

Referring to the species by the initials of their names, a summary of the facts is that we have an individual who does not belong to any species listed HEg, HEr, HH, HN, HS. But, on the other hand, they seem to be related to HEg, using the metaphor of the chain as they would be among some of these species in the evolutionary tree. Is it another missing link then? How a satisfactory explanation could be given?

The conclusion offered by scientists from Atapuerca was to propose an abductive solution as an explanation, which was published in *Science* in 1997 (Arsuaga 2001). A new hypothesis was proposed, according to which there is a new species, hitherto unknown, called *Homo Antecessor*, common to the evolutionary line that leads to HN species in Europe and the evolutionary line that leads to modern populations in Africa, i.e., to HS. In this way the facts were adjusted to the prevailing theory at that time. Nevertheless this does not preclude the proposed hypothesis to be revised if new data or surprisingly new facts were discovered, or if that were contradictory with such hypothesis, leaving it, or keeping some assumptions that are contradictory.

As it has been repeated, abduction is not the only way of inference, but the emergence of something new, or, where appropriate, an anomaly, launches an abductive inference process that concludes with a hypothesis extending (or changing) the initial theory. It is, in short, the methodology of this discipline, which also found, *mutatis mutandis*, to the development of a proto-language, a task that uses comparative methods and it is helped by linguistics, paleontology, archeology, history, etc.

Just as an example, nowadays spoken European languages (with some exceptions, such as Hungarian, Finnish and Euskera) are descendants of Indo-European, a proto-language reconstructed with an evolutionary sense, as noted in this paragraph (Rodríguez-Adrados 2008, 53).

> ... the oldest Indo-European (IE I) we can reconstruct consisted of a series of roots that acquires functions and semantic and grammatical values by means of word order, stress, bonded items, derivation, composition, etc. Only after certain period endings were created with grammatical value (in IE II) and later still with themes with grammatical value. This happened when IE III was created and, within it, each of the two sectors, A and B.[4]

However, deduction can also be used to formulation of hypotheses in the context of discovery. This is the case when some of the sentences involved in the abductive process were obtained by deduction. A deductive conclusion could be obtained and, finally, proposed as a hypothesis because some of the premises have been borrowed, so to speak, in a different context from that one in which the hypothesis is proposed. Therefore it has a theoretical nature, since it is used not only as a specific theory, but all the available theoretical basis, hence it is considered deductive inference from the totality of knowledge, and it is called the *theoretical preduction* (Rivadulla 2007, 208).

---

[4]"... el más antiguo indoeuropeo que podemos reconstruir (el IE I) consistía en una serie de raíces que adquirían funciones y valores semánticos y gramaticales mediante el orden de las palabras, el acento, elementos aglutinados, derivación, composición, etc. Solo en fecha posterior se crearon desinencias con valor gramatical (en el IE II); y en una posterior todavía, temas con valor gramatical. Esto sucedió cuando se creó el IE III y, dentro de él, cada uno de sus dos sectores, A y B," in the original.

This notion is intended for physics and becomes an application of the *hypothetical-deductive* method in the context of discovery, although it requires that combinations and substitutions made in the relevant equations were consistent with dimensional analysis. For other disciplines an equivalent principle should be required, a principle of combination that legitimizes the use of specific assumptions of a different field in another one.

Whatever the case may be, a question immediately arises, how can we consider that a deductive conclusion is revisable? Assuming a principle of *translational assumptions* in the task of building a proto-language, we consider that we have established certain features of IE III and a new language is discovered, belonging to Indo-European family, where such features are not, then we can think that features would have been lost, which would be a hypothesis deduced from the full knowledge driven by some linguists.

The Hittite (an Indo-European language, so classified by certain characteristics, which was studied when a single IE was considered) is single-case, and this system is the basis of the poli-theme in IE III (Rodríguez-Adrados 2008), in which, however, there are traces of the previous single-case. The IE II proposal for a single theme, which would be deductive consequence of a number of premises (perhaps general laws obtained from river names, place names, archaeological references, etc.), allows us to consider an evolution from IE II to IE III, which facilitates even more to fit the Hittite into an Indo-European family. Thus (it is only a simplification, aiming to clarify the meaning of the notion), it has different "deductive" hypothesis, which have more or less impact on the tasks of interpreting the emerging data.

6.4   Logical Treatment of Abduction

In van Fraassen (2011a) it is said that logic has made it possible to refine philosophical problems that arise in sciences. In fact, the problem of definability and the problem of old evidence are treated by means of logical tools, namely Löwenheim-Skolem and Beth's theorems, on the one hand, and the role of conditional inferences (and a discussion about the importance of a deduction theorem) and probabilistic terms, on the other hand. Despite such specific problems and theorems, let us see how that general idea of using logical tools could be implemented when the problem to be tackled is the abduction.

For a logical study of abduction we should start from notions of classical logic and its applications. Since in many disciplines we use specialized languages, we shall consider a formal language L with the propositional connectives and usual quantifiers. Instead of talking about "propositions," which can be represented by formulas of L, we shall refer to sentences of L, i.e., formulas without free variables. The semantics is taken in terms of the model theory: each L-structure, L-model, or, for simplicity, model, is defined as $M = (D, I)$, where D is a nonempty set, while I represents the interpretation function, which maps elements of the domain to the (individual) constant of L and predicates and relations defined[5] in D to the predicative symbols (of each arity) of L. For an atomic sentence, for example, Pa, we says that M *satisfies* Pa (or that M is a model of Pa, or Pa is true in M) if and only if — hereafter iff — $I(a)$ belongs to $I(P)$, and for a dyadic predicate, for example, Rab, M *satisfies* Rab iff the pair $(I(a), I(b))$ belongs to $I(R)$. Similar when the predicate is of any arity.

In general, according to the principles of composition of formulas with the logical constants (the negation of a sentence A is true in M iff A is false in M, and so on), we say that for every sentence A of L, M satisfies A, or M does not satisfy A. If M satisfies all sentences of a set S, we say for short that M satisfies S (or it is a model of S). M satisfies A, abbreviated as $M \models A$ (analogous respect to a set S).

Given a set S of sentences of L and a sentence A of L, A is a *logical consequence* of S (or S *logically entails or implies* A), in classical sense, iff for every model M, if $M \models S$, then $M \models A$. Symbolically we represent that as $S \models A$. Instead, you can define *all the consequences* of S as the set of sentences of L such that every model of S is also a model of them, i.e. $Cn(S) = \{A: S \models A\}$. They are interchangeable concepts, because for each sentence A of L, $S \models A$ iff A belongs to $Cn(S)$.

If A is not a logical consequence of S, we say that A is (semantically) *independent* of S. Naturally, the pair (S, A) is a *valid* argument (in classical sense) iff $S \models A$. Each argument of L is a formal counterpart of an argument from a fragment of ordinary language that constitutes the specialized language of the corresponding discipline.

This relationship of logical consequence (or entailment) has characteristic properties, independent of the formal language used: the so-called *structural rules*, namely, *reflexivity*, according to which for every set of sentences S it is verified that $S \models A$. *Monotonicity:* if S is a subset of S' and $S \models A$, then $S' \models A$. *Transitivity*, if $S \models A$ and $A \models B$, then $S \models B$. Likewise, there is *compactness*, according to which, if $S \models A$, then there is a finite subset S^* of S, such that $S^* \models A$. Finally, *uniform substitution* of non-logical terms with certain restrictions.

The classical calculi are defined as formal systems that try to account for the logical consequence relation. Given a set of sentences S and a sentence A, when a calculus X is appropriated (or sound) and complete, then "A is proved in the calculi X from S," symbolically $S \vdash A$, is equivalent to say that "$S \models A$." That is to say, in such case it is verified that $S \vdash A$ iff $S \models A$.

The classical logical treatment of abduction has been often studied in terms of deductive explanation, in the sense that from a set of statements (background theory), and a sentence logically independent of this theory, a solution could be achieved in a way that such solution and the theory logically imply the sentence. In Aliseda (2006), the best representative of the AKM-Model, she establishes several types of abduction, according to certain requirements, as it is summarized below. Let T be a background theory (a set of sentences in the corresponding language) and a sentence F (of such language), then (T, F) is an abductive problem if T does

---

[5] A predicate defined in D is a subset of elements of D, a m-adic relation defined on D is a set of m-tuples of elements of D.

not entail logically F, and a sentence A is an abductive solution, one of the classes that are indicated in italics, if the following clauses are verified

1. *Plain*, $T, A \models F$
2. *Consistent*. Besides the requirement of plain abduction, A is consistent with T
3. *Explanatory*. It is the plain abduction that verifies that F is not a logical consequence of A (it is not verified that $A \models F$)
4. *Minimal*. If it is plain and A is the minimal solution, i.e., A is different from the conditional $T \rightarrow F$ ("$\rightarrow$" represents the material conditional)
5. *Preferential*. If it is plain and it is the best solution according to some preferential criterion

The nomological-deductive model of scientific explanation (Hempel-Oppenheim 1948) can be presented in abductive terms. Basically, the question that arises is that given a scientific law L (or a set of laws), with the antecedent conditions C, the fact F is explained as a deductive conclusion. F is called *explanans*, while the *explanandum* consists of L and C. The relationship between *explanans* and *explanandum* must be a logical deduction, in this case in the sense of classical logic, that is to say $L, C \vdash F$.

As the systems of classical logic (first order) are complete and sound, both mentioned relations, $\models$ and $\vdash$, are equivalent. In abductive terms, the scientific explanation can be seen as an abductive problem (L, F), the law and the fact to be explained, and certain process of abductive inference in a way that the antecedent conditions C would be obtained, so $L, C \vdash F$, (equivalently, $L, C \models F$), and at least conditions for a consistent explanatory abduction should be accomplished.

Now the van Fraassen's critical attitude can be mentioned. The scientific explanation as a form of abduction is studied in Aliseda (2006). Is it legitimated? This nomological-deductive model has been extensively discussed and a general conclusion may be its inadequacy, since the idea of scientific explanation has aspects that are not captured by such model. For example, there must be syntactical characteristics that should be accomplished by L and C. However, Aliseda's proposal is a simplification to point out that the covering law model could force the issue of antecedent conditions to be searched for by means of an abductive process. In spite of the nature of scientific explanation, if laws and facts arise as abductive problem, as it is implicit, then how antecedent conditions should be fitted is more a logical problem than a problem of the specific scientific practice.

When an abductive problem is arisen, the fact is surprising in so far as it is a sentence of the language that one would expect to be entailed by the background theory, but it is not. However, given the theory T and fact F, if this is not a logical consequence of the former, then there are two possible cases, namely

1. $\neg F$ is not a logical consequence of T, then F is an *abductive novelty*
2. T logically implies $\neg F$, i.e. $T \models \neg F$, then it is an *abductive anomaly*

In the second circumstance, it is not enough to look for a new sentence A, because whatever it is, T together with A do not entail $\neg F$, given the monotonicity of the classical logical consequence relation. Then F and (T, A) are independent. In this case F should be rejected, or the theory must be revised, which leads to the theories of epistemic change, as the so called AGM model, in accordance with the initial of the authors (Alchourrón et al. 1985).

In short, given a consistent theory T, closed under logical consequence (if $T \models A$, then A belongs to T), it is a belief state for an epistemic subject, so that when we have a new sentence F, there can be three epistemic attitudes, namely *acceptance*, which means inclusion in T, *rejection*, so that $\neg F$ is included in T, and, finally, *uncertainty*, according to which neither F nor $\neg F$ would be incorporated to T. Hence, three transactions are considered in AGM to characterize the change of belief: *expansion, revision*, which are removed from the theory of judgments that would lead to an inconsistency (with new beliefs to be taken); and *contraction*, according to which some sentences of the theory (also, their consequences) must be rejected without adding any new one.

The novelty and abductive anomaly correspond with attitudes epistemic of uncertainty and rejection of AGM and lead to other two operations of epistemic change in the abductive model (Aliseda 2006). Let the theory T be and the fact F as the abductive trigger,

1. *Abductive expansion*. If F is an abductive novelty, not a logical consequence of T, we can obtain an abductive solution A (consistent and explanatory), then it is added to T
2. *Abductive revision*. If F is an anomaly, then T is reviewed and T' is obtained, so that in T' those statements of T that entailed $\neg F$ are no longer, the solution A is searched for in a way that $T', A \models F$. Thus, this process includes both a contraction and an expansion.

Before concluding this overview, we recall briefly that in classical logic there are logical calculi to capture the relation of logical consequence, as deductive-natural Gentzen's calculus, resolution, semantic tableaux, etc. If $\vdash$ represents one of such calculi, as we have said before, it is sound and complete iff for all set of sentences S and the sentence A, it is verified that $S \models A$ iff $S \vdash A$. Hence, everything said above on abduction by appealing to the logical consequence relation can be expressed in terms of the theorems of the calculus, which verifies the structural rules, i.e., reflexivity, monotony and transitivity; it also verifies uniform substitution. So, given an abductive problem (T,F), A would be a solution if $T, A \vdash F$. If the language is a higher order one, then the corresponding calculus can not be complete in standard sense. It is more common, however, the use of a first-order language, while the use of second order (or higher) is acceptable as long as it properly modifies its semantics.

Some classical calculi have been used to search for solutions to abductive problems. This is the case of *semantic tableaux*, whose method consists of decomposing each complex sentence that belongs to an initial set (the *root*), in accordance with its rules of semantic composition. All branches are opened or closed. A branch is closed when it contains two complementary literals,[6] in another case, the branch is open. A tableau is open, when some branch is open, or closed, when all branches are closed.

The main property of semantic tableaux is the following one. Given a finite set of sentences S and a sentence B, $T \models B$ iff the tableau whose root is formed by T and $\neg B$ is closed. This is equivalent, given sound and completeness, to $T \vdash B$ iff such tableau is closed. If (T,F) is an abductive problem, the tableau with root $(T, \neg F)$ is constructed and, if it is open, then any sentence A that closes any open branch, provided A is added, would be an abductive solution, since the tableau constructed with root $(T, A, \neg F)$ would be closed, so that $T, A, \vdash F$.

In short, the classical model of abduction considers three basic parameters, namely a *background theory* T, the abductive *trigger* (the surprising fact), and the *inferential* parameter, the logical consequence relation $\models$, or the calculus $\vdash$. With semantics tableaux, according to the main property, we can define an abductive calculus: T, F *abduces* A iff the tableau with $(T, \neg F, A)$ as root is closed. Now *abduces* refers an abductive inference, which does not verify the above structural rules (reflexivity, monotony, transitivity), submitted to certain restrictions. Some mathematical logicians doubt whether a calculus without reflexivity and transitivity is a logical calculus actually, though certain algorithm (implemented in SW-Prolog), as showed in Soler and Nepomuceno (2006), is possible.

6.5 Explanatory Models of Modal Character

Some problems turn up when calculi are used to capture abduction. In the case of semantic tableaux, a propositional calculus is sound and complete, though the propositional logic is not much expressive, while classical first order logic is sufficiently expressive to represent many scientific laws. However the semantic tableaux method is semi-decidable, in fact sometimes there may be infinite branches. Procedure has been refined by modifying certain rules to get branches finite from formulas of the class that have the so-called *finite model property*, i.e., those formulas that if a model satisfies them, then there is also a finite model that satisfies such formulas. So the class of formulas effectively treatable by this method can be extended, but the decision problem is one that cannot be definitely solved. Moreover, this model does not give a solution, in principle, when a change of logic is necessary.

Given the limitation of some logics, is there any problem to apply logical tools in that topic? In van Fraassen (2011a), in section 3 Conclusion, it is said

> What we can conclude on a positive note is that attention to logic made it possible to formulate rigorous criteria of adequacy for any proposed solutions. In that way, it was possible to rule out certain solutions that were proposed, and thus clear the way for more nuanced and more sophisticated approaches to those problems.

---

[6] A literal is an atomic sentence p, which is positive, or the negation of an atomic sentence $\neg p$, which is negative. p is complementary with $\neg p$, and vice versa.

This attention does not imply applications of classical logic only. These applications may be seen as the starting point to develop applications of other logics, the so called non classical logics, which contain extensions of classical logic and alternatives. In this way, the semantic tableaux method may suggest a change of logic. Let us see a simple example (at propositional level) about that. Suppose $T = \{p \rightarrow q, q \rightarrow r, \neg q\}$ and r as the fact to be explained, the tableau with root T and $\neg r$ provides three branches, one open and two closed:

$\{p \rightarrow q, q \rightarrow r, \neg q, \neg r, \neg p, \neg q\}$, which is open
$\{p \rightarrow q, q \rightarrow r, \neg q, \neg r, \neg p, r\}$, which is closed ($r$ and $\neg r$ are in the branch)
$\{p \rightarrow q, q \rightarrow r, \neg q, \neg r, q\}$, which is closed ($q$ and $\neg q$ are in the branch)

as it can be seen by developing the tableau as usual (# is the sign of closed branch):

$$p \rightarrow q$$
$$q \rightarrow r$$
$$\neg q$$
$$\neg r$$

Literals which close the open branch are p, q and r, the latter should be discarded, it would be the trivial solution, since r is to be explained, but any of the other two literals added to T, even providing an explanation, produce inconsistent extensions of T. Should we accept the principle of explosion[7] as an inferential rule when two contrary statements live together into a given theory? Does this not suggest the convenience of a change of logic, as, for example, the adoption of a paraconsistent logic? In this line, to face the problem of inconsistencies in the development of a scientific discipline, applications of certain logics have been investigated, as the *adaptive logic* (Batens 2007) and the *logic of formal inconsistency* (Carnielli 2006).

There is no problem on logical tools after all. In the classical model of abduction, as we have studied, new sentences have been searched for, but in the history of science we can see certain problems whose solution is not obtained by mere extension of the theory. An example is the transition from classical mechanics to quantum mechanics, which can not be explained by reducing it to a simple addition of new statements to the new mechanics, or the addition of new physical postulates, or a reduction of the previous ones, but adopting a new inferential point of view,

---

[7] Given two sentences A and B, this principle (Pseudo-Scotus's principle) says that $A, \neg A \vdash B$.

with new ways of reasoning about the discussed phenomena. In the last resort, a change of logic.

The use of semantic tableaux, adapted to modal logic, therefore with additional indexes (worlds) to the nodes of the branches of the tableau (a graph, actually), shows that certain formulas in a normal system lead to a close if another modal logic is adopted, for example, by adopting **T** instead of **K**, or **S5** instead of **S4**. On the other hand, a reduction of the AGM theory of belief revision, above mentioned, could be seen as an impoverishment of its dynamic aspects.

Precisely, modal and multimodal logics are adopted to solve problems that the classical model does not tackle in a satisfactory way. van Fraassen (2011b) discusses the sense of Thomason's paradox for belief, connected with Moore's paradox. The crucial point is the existence of a belief predicate that the agent applies to sentences of which it can be said that the agent assents their semantic content. van Fraassen says

> If a person assents to 'A,' then s/he must assent to 'I assent to 'A" on pain of pragmatic inconsistency. But that fact, which is surely correct regardless of what sentence A is, does not mean that s/he must assent to all conditionals of the form (If A then I assent to 'A') nor their converse (van Fraassen 2011b, 21).

This could be considered in the deduction-theorem landscape, which is a controversial thing in modal logics: in general, given a belief operator B and any sentence A, from $A \models BA$ we cannot infer $\models A \rightarrow BA$. For van Fraassen A brings along with it BA, but this is not exactly the classical entailment, because of which he proposes a distinction between consequence on pain of pragmatic incoherence and consequence on pain of classical logical inconsistency. From that subtle distinction it could be seen how deduction theorem cannot be valid when entailment relation is taken in a modal sense. Whatever the case may be, all of that incidentally shows the necessity of considering non classical consequence relations in order to apply logical tools in epistemology.

A dynamic perspective, as we said above, should be adopted. Then modal logic (multi-modal logic, to tell the truth) becomes the right tool. In this line, let us briefly see a multimodal proposal of Bonano (2005), based on a language with three modal operators:

1. B_0A states that at the beginning (time 0) the epistemic agent believes that A (a formula for modal formal language)
2. IA represents that (between time 0 and time 1) the epistemic agent is informed that A
3. B_1A says that in time 1 (after the revision of their belief in the light of information received), the epistemic agent thinks A

For each one of such operators, in the corresponding possible worlds semantics, there is an accessibility relation. A restriction must be required: the information operator does not fall under the scope of the two operators of belief, what prevents strange expressions like $B_0A \& I\neg B_0A$, which would mean that the epistemic agent initially believes A and it is reported later that he does not believe A.

Accessibility relations for the two operators of belief, B_0 and B_1, determine the set of worlds that agents consider possible in a given world. The information I determines the set of states of information that are compatible with the information received. That is, for the worlds v and w we have

1. $[R(B_0)]vw$ indicates that in the world v the epistemic agent initially considers w as possible
2. $[R(I)]vw$ indicates that in the world v, the epistemic agent informed (between 0 and 1), he considers w as possible
3. $[R(B_1)]vw$ indicates that in the world v, the agent informed in the time 1, he considers w as possible

The *qualitative Bayes' rule* (in probability terms, to which we do not pay attention in this work) is considered as equivalent to a *conservative principle* for revision of beliefs, according to which, if the given information does not contain surprise, then all previous beliefs should be maintained and any new belief should be deducible from the ancient one and the information (Bonano 2005). The evaluation of the operators, after establishing an assignment of truth values to formulas in each world, $V(A,w)$ are defined

1. $V(B_0A,w)=1$ iff for every world v such that $[R(B_0)]wv$, $V(A,v)=1$
2. $V(B_1A,w)=1$ iff for every world v such that $[R(B_1)]wv$, $V(A,v)=1$
3. $V(IA,w)=1$ iff for every world v such that $[R(I)]wv$, $V(A,v)=1$ and there is not another world where A is true

Three specific axioms are given to account for the conservative principle mentioned. These are

1. *Qualified acceptance*: $(IA\&\neg B_0\neg A)\to B_1A$, it states that if the individual is informed that A and initially she thought it was possible, then she accepts it in the revised believes
2. *Persistence*: $(IA\&\neg B_0\neg A)\to(B_0C\to B_1C)$ says that if the individual receives information that contradicts her initial beliefs, then she continues believing what she believed before
3. *Minimality*: $(IA\to B_1C)\to B_0(A\to C)$ says that belief should be reviewed minimally: no belief should be added unless they are implied by the old one and new information

Finally, it incorporates a last modal operator U, that is thought, for a formula A, as A is true in a global sense, so that $V(UA,w)=1$ iff the set of worlds in which A is true is the total set of worlds (symbolically $\|A\|=W$). The formal system consists of the axiom **K** (basic modal axiom of normality, distributivity of modal operators with respect to $\to$) for B_0, B_1 and U and **S5** axioms for U, but I is not a normal operator.

Thus we have a model that attempts to explore the dynamics of knowledge and information beyond the AGM theory. In short, now new information causes a revision of the old beliefs in a manner, which does not coincide exactly with that of such theory, and offers new perspectives to tackle the problem of logical omniscience (remember that in AGM the body of knowledge of the epistemic agent is closed under logical consequence).

Obviously, the classical model of abduction can not be subsumed under this proposal. Nevertheless, some aspects could be reexamined. Let T be a background theory and C a novelty. A represents an abductive solution when the information of A implies that certain formula holds. In short, it is verified $IA \rightarrow B_1(A \rightarrow (T \rightarrow C))$. If IA is achieved, according to conditions of abductive problems, $A \rightarrow (T \rightarrow C)$ should be accepted (consistency is assumed), so that $B_1(A \rightarrow (T \rightarrow C))$, then $B_1A \rightarrow B_1(T \rightarrow C)$, but B_1A is a consequence of IA in this system, then $B_1(T \rightarrow C)$, because of which $B_1T \rightarrow B_1C$, since B_1 is a normal operator. B_1T as T is the background theory, so that B_1C, and C must be believed in the time 1, after the abductive process.

Finally we present an *explanatory model* for abduction (Soler et al. 2012), also based on modal logic, with a semantics where each world has associated a logic, a semantic of possible logics. The logic of a world may be different from the logic of another different world, which means that conditions for a formula B to be entailed by a set of statements S may be different from one world to another, because any world could be governed by the classical logical consequence, another by a relevant logical consequence, another by a non-monotonic system, etc. Formally, an explanatory model is defined as $M = (L, W, r, R, k)$, where

- L is a formal language
- W is the set, nonempty, and at most countable, of worlds
- r is a nonempty set of acceptable logics
- R is the accessibility relation between worlds
- k is a mapping such that if w is a world, then $k(w)$ is a logic of r.

We interpret logics of r as sets of formulas closed under logical consequence (but not necessarily in the classical sense). k determines the set of formulas that are true in each world. We can assume that R is monotonic in k, i.e., for any worlds v, w, if $Rk(v),k(w)$, then $k(v) \subseteq k(w)$ — $\subseteq$ expresses that the former set is a subset of the second one — but we do not take it as an essential restriction, because it might be useful, for example, to model revision of theories instead of a mere expansion. In addition besides the initial language L a meta-language L^* will be used, in order to reason about changes of theories, so that L^* contains all formulas of L and is closed under negation, conjunction, disjunction, and implication, symbolized $\sim$, $\cdot$, o, $=>$, respectively. Such symbols are different from $\neg$, &, v, $\rightarrow$, the connectives of L (we do not deal with quantifiers for short). L^* contains modal operators (+L), (−L), which represent *ascending* and *descending* necessity, and the corresponding dual operators, (+M) and (−M), which can express *up* and *down* possibility, respectively. Thus, if A and B are formulas of L^*, then $\sim A$, $A \cdot B$, AoB, $A => B$, (+L)A, (−L)A, (+M)A, (−M)A are also formulas of L^*.

The semantics of L^* is defined as follows. To each formula A of L^* it is assigned a set of worlds $\|A\|$, which consists of those worlds in which the formula is valid, so we set either w belongs to $\|A\|$, or $M, w \models A$:

1. $\|\sim A\|$ is the complementary of $\|A\|$, i.e. $W - \|A\|$
2. $\|A \cdot B\|$ contains the worlds that belong to both $\|A\|$ and $\|B\|$, the intersection of $\|A\|$ and $\|B\|$

3. $\|A\ o\ B\|$ contains the worlds that belong to $\|A\|$ or to $\|B\|$, i.e., the union of $\|A\|$ and $\|B\|$

4. $\|A => B\|$ contains the union of the worlds that belong to $\|\sim A\|$ or to $\|B\|$

5. $\|(+L)A\|$ is the set of worlds w' such that, if w belongs to $\|A\|$ and $Rww,'$ then w' belongs to $\|(+L)A\|$, i.e., the set of worlds that are accessible from the worlds that belong to $\|A\|$

6. $\|(-L)A\|$ is the set of worlds w' such that if w belongs to $\|A\|$ and $Rw'w$, then w belongs to $\|(-L)A\|$, i.e., the set of worlds from which the worlds that belong to $\|A\|$ are accessible

As a concrete example, consider the following explanatory model M, in which the set of worlds $W = \{a, b, c, d, e, f\}$, R is a reflexive relation that contains the pairs (x,x) for every x of W, and (a, c), (a, d), (b, d), (c, e), (d, e), (d, f), so that the atomic formula p is true in the worlds a, c, e and false in the other ones, so that $\|p\| = \{a, c, e\}$. Then, it is verified that

1. $M, b \models (+L)\neg p \cdot \sim (+L)(+L)\neg p$. First, in worlds that are accessible from b (only the world d) $\neg p$ is true. In the worlds that are accessible from the worlds where $(+L)\neg p$ is true, $\neg p$ should be true, but e is accessible from d and $\neg p$ is not true in e, so that $(+L)(+L)\neg p$ is false and, consequently, $\sim (+L)(+L)\neg p$ is true, so that $(+L)\neg p \cdot \sim (+L)(+L)\neg p$ is true. It should be noted that $\neg$ represents the negation in L, while $\sim$ expresses the negation in L^*.

2. $M, c \models p \cdot (+L)p \cdot (-L)p$. In fact, c belongs to $\|p\|$, then $M, c \models p$, and e is the only world that is accessible from c, and e belongs to $\|p\|$, and by definition, e belongs to $\|(+L)p\|$, then $M, c \models (+L)p$. On the other hand, a is the only world of $\|p\|$ from which there are accessible worlds of $\|p\|$ (a, and e) so that a is also in $\|(-L)p\|$, then $M, c \models (-L)p$. Therefore the conjunction of the three formulas is satisfied, that is to say $M, c \models p \cdot (+L)p \cdot (-L)p$

We can now formally define an abductive problem in M as the pair (w, F), a world w and a formula F, which verifies

1. $M, w \models (+M)F$
2. $M, w \models (+M)\sim F$

which expresses the idea that the formula F, for the world w, is an abductive problem (in a given model), provided there is a world accessible from w in which F is true, and there is also a world accessible from w in which F is false ($\sim F$ true). A logic may be associated to a world, then we can extend the interpretation. When you get an abductive solution, we move from a world w' such that $Rk(w)$ $k(w')$ and F is true in $k(w')$. We can ask whether this new logic $k(w')$ is different from the logic $k(w)$. It could be that $k(w')$ includes new postulates (as in the classic abduction), but perhaps we reach a different style of reasoning, as it would be in a change of paradigm, for example. About the classical abductive solution, this can also be represented in this framework: a formula A is a solution if it is verified

1. $M, w \models (+L)(A \Rightarrow F)$: A in the current logic $k(w)$ explains F
2. $M, w \models (+M)A$: A is admissible in $k(w)$ (in the current logic, that is, on the background theory that is considered in the world w with the corresponding logic)
3. $M, w \models (-M)(+M)(A \cdot {\sim} F)$, which establishes that A is not sufficient by itself to explain F

Thus we have a consistent explanatory abduction. By taking into account the five given characteristics to define abductive problem and abductive solution, we have a formula such that if it is satisfied by M, w, then $(k(w), F)$ is an abductive problem and A is a consistent abductive solution. Such formula is as follows

$$(+M)F \cdot (+M) \sim F \cdot (+L)(A \Rightarrow F) \cdot (+M)A \cdot (-M)(+M)(A \cdot \sim F)$$

Classical abduction can be explained in these terms. Consider $M = (L, W, r, R, k)$, in a way that for each w in W, $k(w)$ is a classical calculus modulo of a set of formulas Tw (which represents the background theory T in the world w), $\vdash$-(Tw) in symbols. For each set of formulas S and a formula A, it is defined $S \vdash$-$(Tw)A$ iff $S, T \vdash A$, which states that A follows from S modulo Tw. Now the accessibility relation R is defined as inclusion: for worlds $w, w,'$ $Rk(w)k(w')$ iff $T(w) \subseteq T(w')$. But inclusion is *reflexive*, and *transitive*, so this model satisfies the axioms of the modal system **S4**, that is, for formulas A and B of L^*, the axioms

- **K**: $(+L)(A \Rightarrow B) \Rightarrow ((+L)A \Rightarrow (+L)B)$
- **T**: $(+L)A \Rightarrow A$
- **4**: $(+L)A \Rightarrow (+L)(+L)A$

Standard rules are taken: *modus ponens*, from $\vdash A \rightarrow B$ and $\vdash A$ infer $\vdash B$, and necessitation, from $\vdash A$, infer $\vdash (+L)A$. These axioms can also be expressed with the other necessity operator $(-L)$, and there are two bridge axioms, between operators, namely

- **M1**: $A \Rightarrow (+L)(-M)A$. If A is valid, then in all accessible worlds (from the current one), there is at least one world (the current one) from which it has been accessed in which A is valid.
- **M2**: $A \Rightarrow (-L)(+M)A$. If A is valid, then in all the worlds from which the current is accessible, there is at least one that is accessible on the one in which A is valid.

Besides the Bonano's perspective (and the last multimodal proposal), the dynamic point of view can be developed by means of the direct use of dynamic epistemic logic, so to say. The basic idea is to define abductive problem, and abductive solution, in terms of knowledge and belief of an epistemic agent, then abductive solution can be added to the information of such agent. Let us shortly look into that to finish this overview about the logical treatment of abduction.

In order to represent the agent's knowledge and belief, the formal language (a propositional one) incorporates two (existential) modalities that are interpreted in the plausibility framework, as given in van Ditmarsch et al. (2007). Then abductive reasoning can be described within the plausibility models framework. In fact, the trigger of an abductive problem is now an observation (an epistemic action) and two phases should be considered: one before the epistemic action and another after it. On the other hand, the action will turn a given formula into an abductive problem if the formula is known after the action but was not known before (Nepomuceno-Fernández et al. 2013). For formulas *A* and *B*, *KA* and *[A!]B* represent 'the agent knows *A*' and 'after observing *A*, *B* is the case,' respectively.

Abductive problem and solution are defined as follows. Let (*M*,*w*) be a pointed plausibility model, and consider (*MA!*,*w*), the pointed model that results from observing *A* at (*M*,*w*). A formula *B* is an abductive problem at (*MA!*,*w*) iff it is known now but it was not known before the observation, i.e., iff (*MA!*,*w*) satisfies *KB* (where K is the knowledge operator) and (*M*,*w*) satisfies ¬*KB*. A formula *B* can become an abductive problem at (*M*,*w*) iff it is not known and it will be known after observing *A*, that is to say, iff (*M*,*w*) satisfies ¬*KB&[A!]KB*. So, this definition is relative to an agent's information at some given stage (what the pointed model (*M*,*w*) represents).

When *B* is an abductive problem at (*MA!*,*w*), then *C* is an abductive solution iff the agent knew before the observation that *C* implies *B*, that is to say, iff (*M*,*w*) satisfies *K*(*C*→*B*). In this case *C* is not known before the observation that triggers the abductive problem, since if *C* were known, we would have *KC* and *K*(*C*→*B*), because of which *KB*.

6.6 Concluding Remarks

The abductive classic model has the classical logic as inferential parameter, but scientific research does not require that the underlying logic has to be such logic. Hence a study of abduction with a non classical logic is quite conceivable. Although **S4** is an extension, strictly speaking, modal logics are taken as non-classical logics. A model as the shown provides some early benefits. On the one hand, the classical model of abduction, as mentioned, is merely a special case of this explanatory model, in which each possible logic is a classical logic modulo a set of formulas (theory). The logic of each world is context sensitive, so to speak, and it represents the underlying logic, while the explanatory model itself, a metalogical system after all, may be a modal logic **S4**.

On the other hand, from the studied applications of logic we could explore more sophisticated applications, for example, to obtain an explanatory framework that will tackle both the AGM theory of epistemic change and some issues of previous multimodal framework for belief revision. The logic of each world can be specific. The fact that in certain world the AGM postulates were verified is not a problem. In fact the semantics of the explanatory model (a meta-model, actually) is independent of the semantics that corresponds to the logic of every possible world. Similarly, the Bonano's proposal may be relevant in worlds, without affecting the characterization of the proposed explanatory model.

Perhaps van Fraassen's suggestion of using more logic in philosophy of science should be taken into account, since scientific practices could require logical tools to achieve their aims. In fact, many lines of researching in modern logic, as belief revision, many agent perspectives, dynamic logics, etc., seem to be relevant to understand a part of the work in philosophy of science. Finally, the use of Bonano's dynamics system, the meta-theoretical point of view of the explanatory model as a modal system **S4**, or the direct dynamic treatment of abduction, can somehow define a universal abductive logic, whose structural description should be investigated.

References

Alchourrón, C., Gärdenfors, P., & Makinson, D. (1985). On the logic of theory change: Partial meet contraction and revision functions. *Journal of Symbolic Logic, 50*, 510–530.

Aliseda, A. (2006). *Abductive reasoning: Logical investigations into discovery and explanation* (Synthese library, Vol. 330). Dordrecht: Springer.

Arsuaga, J. L. (2001). *El enigma de la esfinge*. Barcelona: Plaza y Janés, Círculo de Lectores.

Batens, D. (2007). A universal logic approach to adaptive logics. *Logica Universalis, 1*, 221–242.

Beuchot, M. (2002). *Estudio sobre Peirce y la escolástica* (Cuadernos de Anuario Filosófico, Serie Universitaria 150). Pamplona: Universidad de Navarra.

Bonano, G. (2005). A simple modal logic for believe revision. *Synthese, 147*(2), 193–228.

Carnielli, W. (2006). Surviving abduction. *Logic Journal of IGPL, 14*(2), 237–256.

Gabbay, D., & Woods, J. (2006). *A practical logic of cognitive systems. Vol. 2: The reach of abduction: Insight and trial*. Amsterdam: Elsevier.

Hempel, C. G., & Openheim, P. (1948). Toward a theory of the process of explanation. *Philosophy in Science, 15*(2), 135–175.

Hintikka, J. (1999). What is abduction? The fundamental problem of contemporary epistemology. In J. Hintikka (Ed.), *Inquiry as inquiry: A logic of scientific discovery* (pp. 91–113). Dordrecht: Kluwer Academic Publisher.

Lipton, P. (1996). *Inference to the best explanation*. London: Routledge.

Magnani, L. (2009). *Abductive cognition: The epistemological and eco-cognitive dimensions of hypothetical reasoning* (Vol. 3 of Cognitive system monographs). Dordrecht: Springer.

Martínez-Freire, P. (2010). Observaciones sobre el concepto de abducción. In H. van Ditmarsch, F. Salguero, & F. Soler (Eds.), *Liber Amicorum Ángel Nepomuceno* (pp. 77–84). Sevilla: Fénix Editora.

Nepomuceno-Fernández, A., Soler-Toscano, F., & Velázquez-Quesada, F. R. (2013). An epistemic and dynamic approach to abductive reasoning: Selecting the best explanation. *Logic Journal of the IGPL*. doi:10.1093/jigpal/jzt013.

Peirce, C. S. (1991). *The essential Peirce. Selected philosophical writings* (Vol. I). Bloomington: Indiana University Press.

Peirce, C. S. (1998). *The essential Peirce. Selected philosophical writings* (Vol. II). Bloomington: Indiana University Press.

Psillos, S. (1996). On van Fraassen's critique of abductive reasoning. *The Philosophical Quarterly, 46*(182), 31–47.

Rivadulla, A. (2007). Abductive reasoning, theoretical prediction, and the physical way of dealing with nature. In O. Pombo & A. Gerner (Eds.), *Abduction and the process of scientific discovery* (pp. 199–210). Lisbon: Publidisa.

Rodríguez Adrados, F. (2008). *Historia de las lenguas de Europa*. Madrid: Editorial Gredos.

Soler, F., & Nepomuceno, A. (2006). Tarfa: Tableaux and resolution for finite abduction. In M. Fisher, W. van der Hoek, B. Konev, & A. Lisitsa (Eds.), *Logic in artificial intelligence. 10th European conference, JELIA 2006* (pp. 511–514). Berlin: Springer.

Soler, F., Fernández, D., & Nepomuceno, A. (2012). A modal framework for modelling abductive reasoning. *Logic Journal of IGPL, 20*(2), 438–444.

Thagard, P., & Shelley, C. (1997). Abductive reasoning: Logic, visual thinking, and coherence. In M. L. Dalla Chiara et al. (Eds.), *Logic and scientific methods* (pp. 413–427). Dordrecht: Kluwer Academic Publisher.

van Ditmarsch, H., van der Hoeck, W., & Kooi, B. (2007). *Dynamic epistemic logic*. Dordrecht: Springer.

van Fraassen, B. (1980). *The scientific image*. Oxford: Clarendon.

van Fraassen, B. (1989). *Laws and symmetry*. Oxford: Clarendon.

van Fraassen, B. (2011a). Logic and the philosophy of science. *Journal of the Indian Council of Philosophical Research, 27*, 45–66.

van Fraassen, B. (2011b). Thomason's paradox for belief, and two consequence relations. *Journal of Philosophical Logic, 40*, 15–32.

Woods, J. (2012). Cognitive economics and the logic of abduction. *The Review of Symbolic Logic, 5*(1), 148–161.

Chapter 7
The View *from Within* and the View *from Above*: Looking at van Fraassen's Perrin

Stathis Psillos

Abstract Bas van Fraassen has usefully contrasted two ways to view the relation between theory and measurement: from above and from within. Roughly put, "from above" is the perspective in which we view measurements from the point of view of the finished theory aiming to examine how the measurement is related to the theory. "From within" is the perspective in which we see measurement as a means for the development of the theory. van Fraassen warns us that we need a "synoptic vision," one that combines both perspectives. In this chapter, I argue that this synoptic view can be had *without* forfeiting important conclusions about how theory and experience and observation are related to reality. I make my case by looking in detail into an important episode in which the two views should clearly be in play: Perrin's work on the Brownian motion. This case has been recently studied by van Fraassen too. There are significant elements of disagreement in the ways we look at this case. I argue that Perrin's case shows that it is unreasonable to defend the superiority of the molecular theory *c. 1912* without defending its likely truth. There is an important point of contact with van Fraassen: we both take measurement to be a vehicle of representation. But we disagree on the role of instruments as means for representation. After having presented my own way to bring together the view from within and the view from above in Perrin's case, I take issue with his account of instrument-driven measurement as a case of public hallucinations.

An earlier version of this chapter was presented at the *XVI Jornadas de Filosofía y Metodología actual de la Ciencia*, at the University of A Coruna, Spain in March 2011. Many thanks to Bas van Fraassen and to Valeriano Iranzo for comments and to Wenceslao J. Gonzalez for the kind invitation. Research for this paper has been co-financed by the European Union (European Social Fund – ESF) and Greek national funds through the Operational Program "Education and Lifelong Learning" of the National Strategic Reference Framework (NSRF) – Research Funding Program: THALIS – UOA – APRePoSMA.

S. Psillos (✉)
Department of Philosophy and History of Science, University of Athens,
University Campus, w/n, 15771 Athens, Greece
e-mail: psillos@phs.uoa.gr

Keywords van Fraassen • Perrin • Brownian motion • Representation • Atomism • Confirmation

7.1 Introduction

Bas van Fraassen has usefully contrasted two ways to view the relation between theory and measurement: from above and from within. Roughly put, "from above" is the perspective in which we view measurements from the point of view of the finished theory aiming to examine how the measurement is related to the theory. "From within" is the perspective in which we see measurement as a means for the development of the theory. The first perspective is ahistorical while the second is historical. van Fraassen (2008, 139) warns us that we need a "synoptic vision," one that combines both perspectives. He claims that this synoptic point of view frees us from the illusory search for "a view from nowhere". It tells us how theory and measurement are related without presupposing "an impossible God-like view in which nature and theory and measurement practice are all accessed independently of each other and compared to see how they are related 'in reality'" (2008, 139).

This "synoptic" vision brings to mind — as van Fraassen is clearly aware of — Sellars's project to relate the manifest image with the scientific image of the world. Sellars urged us to bring the two images together in a "stereoscopic" view, which is achieved when "two differing perspectives on a landscape are fused into one coherent experience." But, the reader may recall, this Sellarsian stereoscopic view was not symmetric; nor did it put the two images on equal footing, as it were. Accommodation of the manifest image there will be, but the scientific image retains its primacy: "in the dimension of describing and explaining the world, science is the measure of all things, of what is that it is, and of what is not that it is not."

The view from above and the view from within should certainly be brought together in a stereoscopic view. What van Fraassen has done in his (2008) to bring them together is really illuminating. But, I will argue, this stereoscopic view can be had *without* forfeiting important conclusions about how theory and experience and observation are related to reality.

I will make my case by looking in detail into an important episode in which the two views should clearly be in play: Perrin's work on the Brownian motion. This has been recently looked at by van Fraassen himself (cf. 2009). There are significant elements of disagreement in the ways we look at this case. I will try to bring them out as clearly as I can. I will argue that Perrin's case shows that it is unreasonable to defend the superiority of the molecular theory *c. 1912* without defending its likely truth.

There is an important point of contact with van Fraassen: we both take measurement to be a vehicle of representation (cf. 2008, 91). But we disagree on the role of instruments as means for representation. After having presented my own way to bring together the view from within and the view from above in Perrin's case, and after having contrasted it with van Fraassen's, I will take issue with his account of instrument-driven measurement as a case of public hallucinations.

7.2   Brownian Motion c. 1905

It is widely accepted that between roughly 1908 and 1912, there was a massive shift in the scientific community on the European continent in favour of the atomic conception of matter. It is also widely accepted that Perrin's theoretical and experimental work on the causes of Brownian motion played a major role in this shift. Brownian movement — so-called because it was first identified as such by the botanist Robert Brown — is the incessant and irregular agitation of small particles suspended in a liquid. When Perrin received the Nobel Prize for physics in 1926, it was noted in the presentation speech by Professor C.W. Oseen that he "put a definite end to the long struggle regarding the real existence of molecules."

We can see this shift of opinion in its clearest form in Henri Poincaré's writings from 1900 to 1912. In his address to the 1900 International Congress of Physics in Paris, Poincaré claimed that the atomic hypothesis, viz. the hypothesis that matter has a discontinuous structure, is "indifferent," that is, useful as a device of computation or for providing "concrete images" which help scientists fix their ideas (1900, 10). From this, Poincaré noted, there is no reason to conclude "the real existence of atoms." In 1912, in a lecture delivered at the French Society of Physics, Poincaré famously spoke of the experimental proof of the reality of atoms: "the atoms are no longer a convenient fiction; it seems, so to speak, that we can see them since we know how to count them" (1913, 89). What made the difference for Poincaré was Perrin's experiments on the Brownian motion: the shift of the atomic hypothesis from the status of an indifferent hypothesis to the status of a true description of reality was centred around what Poincaré took it to be its experimental verification by Perrin.

Poincaré's change of stance was by no means untypical. Perhaps the most important shift of opinion occurred in Wilhelm Ostwald, who in his (1896) had vehemently attacked "scientific materialism" (basically the "mechanics of atoms"). Already in 1908, in the preface of the third edition of his *Outlines of General Chemistry* he (1912, vi) noted:

> … [T]he agreement of the Brownian movements with the requirements of the kinetic hypothesis, established by many investigators and most conclusively by J. Perrin, justify the most cautious scientist in now speaking of the experimental proof of the atomic nature of matter. The atomic hypothesis is thus raised to the position of a scientifically well-founded theory, and can claim its place in a textbook intended as an introduction to the present state of out knowledge of General Chemistry.

He then went on to present the atomic hypothesis in relation to the Brownian movement and spoke of "the final proof of the grained or atomistic-molecular nature of matter (…) after a fruitless search during a whole century" (1912, 483–484).

The atomic conception of matter went through many twists and turns. It did face significant scientific anomalies (e.g., the specific heats anomaly) as well as important theoretical successes (e.g., the extension of the kinetic theories of gases to liquids). But it did also face important philosophically-motivated objections. Heinz Post (1968) drew a useful distinction between two types of atomic theory: the "essentially atomic theories," which do allow the determination of Avogadro's number N, and those that do not. The atomic hypothesis entailed that atoms should be countable; hence, the determination of Avogadro's number N (the number of atoms in a gram molecule of a gas) was a key plank in the defence of atomic theory. For, among other things, the determination of this number would allow the determination of other atomic properties, e.g., the size of molecules. There had been many attempts to determine Avogadro's number (cf. Brush 1968), notably by William Thomson (1870), who used four ways to estimate it and who actually declared that the atomic hypothesis thereby received "a high degree of probability." But all this did not really sway the balance in favour of the atomic theory. The opposition's case was still very strong.

Though Brownian movement had received some attention by scientists (cf. Brush 1968), the first to embark on a systematic study of it from the point of view of the atomic conception of matter was the French physicist Léon Gouy. The motion of the suspended granules was "incessant" without being subjected to a visible external cause. In fact, Gouy (1895) showed that a number of initially suggested external causes of the Brownian movement (such as the illumination of the particles by the microscope) could be safely eliminated; and so could a number of initially plausible internal causes (such as convection currents). For Gouy, the kinetic theory of gases (committed as it was to the thesis that the structure of matter is granular) offered a cogent explanation of the phenomenon — the random motion of the Brownian particles was due to the movement of the molecules of the liquid in which they were contained. The extremely active internal agitation of the liquid — which defied its appearance as an immobile body — was the cause of the Brownian motion. This invisible but permanent agitation of the molecules of the liquid explained the incessant and indefinite movements of the Brownian particles. The latter offered us a "feeble image" of the molecular motion (1895, 7). Though Gouy did try to make a case for the claim that the atomic conception of matter — in light of the Brownian motion — did deserve serious attention, he also admitted that there was not yet a way available to develop this explanation in a rigorous and measurable manner.

It might be ironic that Gouy's piece titled "The Brownian movement and the molecular motions" appeared in the same volume of *Revue Générale des Sciences Pures et Appliquées* as Ostwald's "The Failures of Scientific Materialism." Ostwald's attack on atomism was predicated on the fact that the atomic hypothesis was not definite and precise enough. And though it was increasingly accepted that potential explanations of Brownian movement other than that based on the kinetic theory of gases were "untenable," as Poincaré (1906) put it in his address at the St Lewis International Congress of Arts and Science, until the middle of the first decade of the twentieth century there was nothing like a proper theory which unveiled the atomic basis (the quantitative mechanism and the laws) of Brownian movement.

Two important things happened in the first decade of the twentieth century, which as Nye (1976, 266) put it, led to "a completely renovated atomic hypothesis." One was Einstein's (1905) theory of Brownian motion, which provided for an explanatory mechanism of it based on the molecular kinetic theory; the other was Perrin's theoretical and experimental work, which allowed a very accurate determination of Avogadro's number. In Perrin's hand, Avogadro's number became an invariant and indispensable feature in explanations of various phenomena. More importantly, it paved the way for the exact determination and measurement of atomic magnitudes.

I will substantially elaborate on Perrin's achievements looked at both from within and from above. But before this, let me outline the way van Fraassen reads Perrin's achievements.

7.3   van Fraassen's Perrin

The story van Fraassen (2009) narrates aims to render Perrin's achievements accountable from a constructive empiricist point of view, and hence to block the claim that there is a *privileged* realist reading of Perrin's work as being the demonstration of the reality of unobservable molecules. He (2009, 5) takes it that the standard interpretation outlined above constitutes the LORE:

> LORE: until the early 20th century there was insufficient evidence to establish the reality of atoms and molecules, but then Perrin's experimental results on Brownian motion convinced the scientific community to believe that they are real.

He takes it that the LORE is an interpretation of Perrin's achievements and aims to offer an alternative *interpretation*.

When Perrin enters van Fraassen's narrative of the adventures of the atomic conception of matter during the nineteenth century, the stage has already been set in a certain way: the atomic conception of matter was in need of *empirical grounding*. This need was perceived equally forcefully by the friends and foes of the atomic conception and was meant to pose a challenge to the developing theory, viz., to provide (empirical and measurable) links between the theoretical parameters and various empirical phenomena.

van Fraassen renders this idea of empirical grounding more precise by introducing some elements of Hermann Weyl's (1927 [1963]) account of measurement. He takes from Weyl two conditions. Here is how he (2009, 11) puts them:

> *Determinability*: any theoretically significant parameter must be such that there are conditions under which its value can be determined on the basis of measurement.
> *Concordance*, which has two aspects:
>
> – *Theory-Relativity*: this determination can, may, and generally must be made on the basis of the theoretically posited connections.
> – *Uniqueness*: the quantities must be 'uniquely coordinated', there needs to be concordance in the values thus determined by different means.

Roughly put, the two conditions require that for a theory to be empirically grounded, its basic theoretical magnitudes must be amenable to measurement and that the various measurements of the values of these theoretical magnitudes must yield roughly the same result. When Weyl introduced these conditions, he intended them to be conditions of "the correct theory of the course of the world" (1927 [1963], 121). He insisted that the theory becomes empirically testable when theoretical magnitudes are linked "by theoretically posited connections" to empirical data in such a way that the values of these theoretical magnitudes are determined. If there are multiple determinations of these values, they should be concordant with each other, for otherwise the theory would be inconsistent. He added, however, that the demand of concordance "brings the theory in contact with experience."

van Fraassen takes Weyl's conditions to capture the empirical grounding of the theory. In fact, they do a lot more. As Weyl (1927 [1963], 141–142) explains a few pages later, the theory of measurement should be able to address the following question: how is it possible "to determine quantities much more accurately than the differentiating capacity of our senses permits"? What he clearly has in mind is that the theory of measurement should be able to address this question also in the case in which *theoretical* quantities are involved. The theory establishes connections (functional relations) between theoretical magnitude x and various others. Measuring these other magnitudes, the value of x can be determined "more exactly than by direct observation." Insofar as the results of the various measurements are accurate *and* concordant, "the basic theories are confirmed."

It might be thought that Weyl does not take this reference to confirmation seriously. But this would be too quick. A theory whose theoretical parameters can be subjected to the requirements of determinability and concordance is, as Weyl puts it, a well-founded theory. This is not simply an empirically grounded theory; it is also a confirmable theory. A bit further on his text, Weyl (1927 [1963], 185) discussed explicitly the case of atomic theory and — without referring to Perrin specifically but with an explicit mention of the relevance of the Brownian movement — talks about the "golden era of atomic research" and stresses:

> During the last half century it has provided a thorough and brilliant corroboration for the basic tenets of atomism and penetrated into ever deeper layers of the strange atomic world. To begin with, all its methods led with increasing accuracy to the same values of the mass and charge of an electron. Only through this concordance has atomistics become a well-founded physical theory. Gradually indirect methods have been replaced by more and more direct ones. Thus the Brownian motion of small suspended particles demonstrates directly to our senses the presence of a molecular thermic motion. Through ingeniously arranged experiments one has succeeded in isolating the effects of individual atomic events.

There should be little doubt as to how this passage should be interpreted. It might not even be an accident that Weyl uses the very same expression that we have seen Ostwald using, when he came to accept the atomic theory: the theory has been "raised to the position of a scientifically *well-founded theory*" (emphasis added).

It might also be thought that the very fact that the determination of the values of the basic theoretical magnitudes is done *relative* to the theory is problematic. But this need not be so, as van Fraassen himself notes. For one, the relativity of the test to the theory shows, as Weyl in effect noted, the indispensability of theories in making certain measurements available. It is only on the basis of theories and of theoretically posited connections between theoretical magnitudes and empirical magnitudes that "a difference" which is "not manifest to the senses" is established (cf. Weyl 1927 [1963], 142). As we shall see later on, this is achieved masterfully by Perrin when he predicted Avogadro's number based on the assumptions that Brownian particles are, simply, large molecules. The key point here, however, is that the relativity of the test to the theory suggests that the theory is ahead of the measurement in making certain measurements available. For another, it is by no means certain that the test will comply with the theory. Hence, its relativity to the theory does not imply that the test is trivialised. This is already achieved in Clark Glymour (1975) well-known bootstrapping theory of confirmation — which van Fraassen (2009, 12) favourably discusses though he prefers to leave behind confirmation and to keep just the bootstrapping — viz., *the relativity to theory*. In any case, as we shall see later on, Perrin's theoretical model of Avogadro's number left it entirely open whether the actual measurements of the properties of the Brownian particles would confirm the kinetic theory of gases.

Weyl's conditions are very natural constraints on a well-founded theory and they had been introduced already by Ostwald — at least in their essentials.[1] The difficulty with van Fraassen's appropriation of them is not with the conditions themselves. Nor is it with the fact that Perrin did not try to ground empirically the atomic conception of matter — he certainly did. The difficulty is with van Fraassen's contention that the requirement of an empirical grounding of a theory is *an end in itself*. Rather, the empirical grounding — the determination and measurement of the basic theoretical parameters of the theory — is a means for the theory to change cognitive status: from being a mere hypothesis to being (reasonably accepted to be) true. Provided, of course, that one does allow — in principle — this change of status. (I tell this story in detail in my 2011.)

van Fraassen (2009, 19) says:

> but it was [Perrin's] achievement to tie the research that was needed, to complete these efforts at the empirical grounding of the theory, to the study of Brownian motion.

What exactly does that mean? The atomic theory was in the process of theoretical development, aiming to apply it to new domains and phenomena. This development required the addition of further hypotheses, which introduced new theoretical magnitudes related to the properties that molecules would have if they existed. There was then need for the specification of these magnitudes and of finding "stricter and stricter connections" between them and measurable quantities. Given this theoretical development which aims to incorporate new phenomena into the theory, "empirical measurements take on a special significance: their outcomes place

---

[1] See my (2011). Ostwald (1907, 408) introduced an important criterion concerning the cognitive status of hypotheses — one that was destined to show why the atomic hypothesis (as developed and tested by Perrin) could change cognitive status. Ostwald noted that *definiteness* and *measurability* were conditions such that, once met, they could change the cognitive status of a hypothesis.

constraints on what the values of the molecular parameters can be." Insofar as the outcomes of these measurements are strict and uniquely determined, the theory is empirically grounded. What Perrin achieved then, according to van Fraassen, is the strict and unique specification of a certain theoretical parameter that was required for the empirical grounding of the molecular model, viz., Avogadro's number.

In his (2008) van Fraassen ties this point about empirical grounding to his view about empirical adequacy. To be sure, he speaks of Millikan and not of Perrin, but the point is virtually the same, viz., that measurements are required for the specification of certain theoretical parameters, for measurements only can show what the value of the theoretical parameter must be if the theory is not to end up being empirically *inadequate*. Extending what van Fraassen says of Millikan to Perrin, Perrin's achievement was that he "filled a blank" in the atomic conception of matter, viz., Avogadro's number. Seen from within, Perrin's experiments wrote a number into a theoretical blank:

> What I mean is: in this case the experiment shows that unless a certain number (or a number not outside a certain interval) is written in the blank, the theory will become empirically inadequate. For the experiment has shown by actual example that no other number will do; that is the sense in which it has filled in the blank. So regarded, *experimentation is the continuation of theory construction by other means* (2008, 112).

There are a couple of objections to van Fraassen's narrative, which will be substantiated after we had gone through Perrin's work. But here they are in outline. The first is that Avogadro's number N had been calculated in various ways before Perrin's own theoretical account of the Brownian motion and the experimental specification of N. So, the role of Perrin's work (both theoretical and experimental) was *not* to fill a blank in a theory — this was already filled, as it were. The role of Perrin's work was to show that a certain way to calculate N (based on a certain theoretical prediction of it) could provide a *decisive test* in favour of the atomic conception. Indeed, history does not seem to be on van Fraassen's side. Perrin's work on the Brownian motion and the atomic hypothesis was deemed so important that he was invited to address the French Philosophical Society on the 27th of January 1910. Though there is no doubt that Perrin wanted to render this number determinate and precise (as he put it, "we consider this determination as given or as highly probable, if we get similar numbers by radically different methods"), he was adamant that this determination was meant to undermine a reason offered by various scientists for taking a fictionalist stance towards atoms, that is for arguing that it is "as if atoms exist" (1910b, 268). For Perrin, his own work was not just meant to "fill in" a theoretical parameter; "it's the true existence of atoms that we claim to establish." Secondly, Perrin's experiments did not aim to prove that the atomic theory would be empirically inadequate unless N has had a certain value. This was well-known too. Rather, they aimed to show that the theory-led determination and concordance (to use van Fraassen's terminology) of certain theoretical parameters can become so precise that resistance to accepting these parameters as real was no longer rational. He closed his aforementioned address by stressing that it will be difficult to defend "by reasonable arguments a hostile attitude to the molecular hypothesis" (1910b, 281).

In the sequel, we shall look in detail into Perrin's strategy for proving the reality of molecules, aiming to show that van Fraassen's interpretation of it is unwarranted.

7.4   Perrin Revisited: The View from Within

Perrin's first paper on the atomic conception of matter was published in 1901 in *Revue Scientifique* under the title: "The Molecular Hypotheses." In this, he took it that, by and large, the debate about the molecular hypothesis — that is the debate about whether matter is continuous or discontinuous, has had a "uniquely philosophical character" and as such the choice between the two approaches was a matter of "taste." The issue could not yet be dealt with experimentally. Perrin did favour the atomic conception, but he was adamant that even though numerous of its consequences had been experimentally confirmed and even though these did not follow from the alternative hypothesis (of continuity), still "we will not perhaps have the right to say that the molecular hypothesis is true, but we will know at least that it is useful."[2] The atomic hypothesis remained "one of the more powerful tools of research" invented by human reason. Perrin presented the rudiments and the successes of the kinetic theory of gases and stressed that the law of the corresponding states that was established by van der Waals was a "triumph" of the theory. But he did also claim that for the acceptance of the atomic conception as something more than useful, the determination of the number of molecules and of their diameter was required.

Ten years later, Perrin published another paper (1911) in the same journal as his 1901 paper, this time with the title: "The Reality of Molecules." The conclusion of this article was that "the objective reality of molecules" had been demonstrated. What had happened in between?

Perrin's more technical work is collected in his *Brownian Movement and Molecular Reality*, which appeared in French in 1909 and was translated into English in 1910. In this book, Perrin makes almost no methodological remarks, but the key point of his strategy is summed up thus: "Instead of taking this hypothesis [the atomic hypothesis] ready made and seeing how it renders account of the Brownian movement, it appears preferable to me to show that, possibly, it is logically suggested by this phenomenon alone, and this is what I propose to try" (1910a, 7).

Perrin takes it that the atomic hypothesis is a plausible hypothesis, its plausibility being grounded in the fact that, in the end of the day, it is the only serious admissible explanation of Brownian movement. Reviewing the work of Gouy and others, Perrin concurs that several potential causes of the movement can be

---

[2] Perrin did stress, already in 1901, that the molecules of gases are composed of atoms and that the atoms have internal structure.

safely eliminated and that, in particular, it is plausible that the cause of the movement is internal and not external (cf. 1910a, 6). This kind of eliminative approach paves the way for rendering the standard atomic explanation of Brownian movement "by the incessant movements of the molecules of the fluid" the only plausible explanation. This is not enough to render it true or probable. In his already noted address to the French Philosophical Society, Perrin (1910b, 273) was clear that having a hypothesis about the *sense*, as he put it, of the Brownian motion was not enough; what was also required was a phenomenon which could allow him to measure directly Avogadro's number. His ingenious strategy was to show that Brownian movement *is* itself an instance of molecular movement and hence that it obeys the laws of the molecular movement. Hence, it can be used to (a) determine Avogadro's number and (b) to specify the individuating properties of atoms.

His theoretical construction proceeds as follows. Let us suppose we have a uniform emulsion (all granules are identical) in equilibrium, which fills a vertical cylinder of cross section s. Consider a horizontal slice contained between the levels $<h, h+dh>$, where this slice is enclosed between two semi-permeable pistons — they permeable to the molecules of water but impermeable to the granules. Each piston is subjected to osmotic pressure by the impact of the granules it stops. This slice of granules does not fall; hence there must be an equilibrium between the force that tends to move it upwards (viz., the difference of the osmotic pressures) and the force that tends to move it downwards (viz., the total weight of the granules less the buoyancy of the liquid). Having estimated both forces, Perrin arrives at the equation of the distribution of the emulsion

$$2/3W \log(n_0/n) = \varphi(\Delta - \delta)gh \qquad (7.1)$$

where W is the mean granular energy, φ the volume of each granule, Δ its density, δ the density of the intergranular liquid and n and n_0 respectively the concentrations of the granules at the two levels separated by height h. The task then is to measure all magnitudes other than W; hence, to determine W (cf. 1910a, 24).[3]

The equation of distribution describes an exponential law. It shows that the concentration of the granules decreases in an exponential way as a function of the height: the concentration is denser towards the bottom of the cylinder and rarer towards its top. This is exactly what happens with the distribution of the density of air: the barometric pressure decreases exponentially as a function of the height — a fact that was known as Laplace's law. It is then this fact that allows Perrin to justify an important claim he makes, viz., that the mean granular energy W of the particles in Brownian motion is equal to mean molecular energy W'. In other words, he

---

[3] As Perrin (1910a, 24, note) stresses, the equation of distribution of emulsion was arrived at independently — and by different means — by Einstein (1905) and Smoluchowski. What Perrin observed, and they did not, was that Eq. (7.1) could furnish a crucial experiment for the molecular theory of Brownian movement.

argued that the Brownian particles behave as large molecules and hence obey the laws of the gases (see also 1916, 89 and 92, 1910b, 275–276).[4]

It was known that the mean kinetic energy W' of the molecules of a gram-molecule of a gas is a function of Avogadro's number N. It is equal to $(3R/2N)T$, where T is the absolute temperature of the gas and R is the constant of the perfect gases (cf. 1910a, 19). Hence,

$$W' = (3R/2\,N)\,T. \tag{7.2}$$

Perrin relied on van't Hoff's proof that the invariability of energy (viz., that the mean kinetic energy is the same for all gases at the same temperature) holds *also* for the molecules of dilute solutions. But he took a step further. By a "rational leap," as he put it (1910b, 272) he generalised van't Hoff's law to *all* fluids, including emulsions. This means that Eq. (7.2) will hold for single molecules, as well as for bigger particles including specks of dust formed by many big molecules. To be sure, Perrin's generalisation of van't Hoff's law follows from the theorem of the equipartition of energy. But Perrin did not take this path because of the complexity of the proof of this theorem (cf. 1910a, 21). In any case, given this generalisation, he (1910a, 20) could note that

> not only (…) each particle owes its movement to the impacts of the molecules of the liquid, but further (…) the energy maintained by the impacts is on average equal to that of any one of these molecules.

The claim that "the mean energy of translation of a molecule [is] equal to that possessed by the granules of an emulsion" — that is that $W = W'$ — is crucial. It paved the way for calculating the granular energy in terms of molecular magnitudes. Accordingly, Perrin thought that the road was open for an *experimentum crucis*: either $W = W'$ or $W \neq W'$ and given that both W and W' could be calculated, we might have "the right to regard the molecular theory of this movement as established" (1910a, 21). It is in this precise sense that Perrin's testing of the molecular origin of the Brownian movement was far from trivial, despite the relativity of the tests to the kinetic theory of gases. For, exactly as van Fraassen would require, the tests could *fail* the theory.

Being an extremely skilful experimenter, Perrin managed to prepare suitable emulsions of gamboge and mastic, with spherical granules of radius α. (7.1) thus becomes

$$2/3W \log(n_0 / n) = 4/3\pi\alpha^3 (\Delta - \delta) gh. \tag{7.1'}$$

---

[4] In *Les Atomes*, Perrin simplifies matters by presenting right away the exponential law applying directly the gas laws to emulsions (cf. 1916, 90–93). This way to proceed might be more appetising, but it might well obscure the justification of the application of the exponential law to emulsions. As he (1916, 93–94) noted with emphasis, the strategy he followed was "to use the weight of the [Brownian] particle, which is measurable, as an intermediary or connecting link between masses on our usual scale of magnitude and the masses of molecules."

Here again, all magnitudes but W are measurable. Determining the ratio (n_0/n) was quite demanding, but Perrin used the microscope to take instantaneous snap-shots of the emulsion. Determining the value α of the radius was even more demand-ing, but Perrin used three distinct methods to achieve this, one relying on Stokes's equation (capturing the movement of a sphere in a viscous fluid), and two without applying this equation (using, instead, a *camera lucida*). These calculations were in impressive agreement, which led Perrin to conclude, among other things, that the otherwise controversial application of Stokes's equation (because it was meant to apply to continuous fluids) was indeed legitimate.

Perrin was then able to calculate the granular energy W (which is independent of the emulsion chosen). If $W = W'$, (if, that is, the Brownian particles do behave as heavy molecules and hence if the laws of the gases do hold for them too), there is a direct prediction of Avogadro's number N from (7.1') and (7.2):

$$(RT/N)\log(n_0 / n) = 4 / 3\pi\alpha^3 (\Delta - \delta)gh$$

and

$$N = 3RT \log(n_0 / n) / 4\pi\alpha^3 (\Delta - \delta)gh. \tag{7.1''}$$

This prediction could then be compared with already known calculations of N based on the kinetic theory of gases, e.g., that by van der Waals's ($N = 6.10^{23}$). Perrin made a number of experiments and concomitant calculations and the agreement was always impressive. As he (1910a, 46) put it: "[I]t is manifest that these values agree with that which we have foreseen for the molecular energy. The mean departure does not exceed 15 % and the number given by the equation of van der Waals does not allow for this degree of accuracy."

Perrin became immediately convinced that "this agreement can leave no doubt as to the origin of Brownian movement." "[A]t the same time," he said, "*it becomes very difficult to deny the objective reality of molecules.*"

Let's be clear on how exactly Perrin argued. Here (1910b, 277) is how he put the matter in his address to the French Philosophical Society:

> To understand how remarkable [this agreement] is, we must consider that before the experiment, they would have certainly not dared certify that the fall of the concentration would not be negligible (…) and that, against it, they would no more have dared to assert that all grains do not gather in the immediate vicinity of the bottom of the tank. The first possibility led to a null value of N, and the second to an infinite value of N. That, with each emulsion, one is landed, in the immense interval which seemed therefore a priori possible for N, precisely on a value adjacent to the expected number, undoubtedly will not appear the result of chance.

Perrin repeated the same point in his more technical work (cf. 1910a, 46, 1916, 105). What does he say? On the negation of atomic hypothesis there are two options available regarding an emulsion suspended in a *continuous* fluid: either all granules stay at the same level or they fall to the bottom of the tank, depending on the viscosi-ties of the fluid and the emulsion. This would lead to calculations of N being either

0 or infinite. The exponential distribution of the Brownian particles is enough to discredit the hypothesis that matter is continuous. But it also discredits any other hypothesis which would give N any other value significantly different from the predicted one. Hence, on the hypothesis that the value of N could be anywhere between zero and infinity, the probability that the predicted value of N is the specific one measured would be zero; on the contrary, this probability is very high given the kinetic theory of gases as developed by Perrin and applied to the Brownian movement. It is precisely this concordance, Perrin noted, that "cannot be considered as the result of chance." Note well that in this setting Perrin takes the negation of the atomic theory to be any theory which predicted *any* other value (order of magnitude) of N (infinite or finite, and in the latter case either zero or any value significantly different from that predicted by the theory). As Perrin put it: "if one was not guided by the molecular theory, once could expect any set of values between and including zero and infinity" (1910b, 281).

Before we try to view Perrin's achievements *from above*, let us note that he does take another step. He stresses that the determination of Avogadro's number by (7.1″) affords a determination of the properties of molecules that can be calculated on its basis. Moreover, this determination of N is now "*capable of unlimited precision*," since all magnitudes in (7.1″) other than N can be determined "to whatever degree of precision desired." Hence, Perrin went on to calculate N and to conclude that its value is $N = 7 \times 10^{23}$. From this, he calculated the weight and the dimensions of molecules. He also reported on a number of other calculations of Avogadro's number N, including from: the measurement of the coefficient of diffusion; the mobility of ions; the blue colour of the sky (the diffraction of the sunlight by the atmospheric molecules); the charge of ions; radioactive bodies; the infra-red part of the spectrum of the black-body radiation. Though all these calculations were less accurate than his own, Perrin took them to prove molecular reality (cf. 1910a, 90), since they are in considerable agreement, showing that this number is "essentially invariant" (1910a, 74).[5]

Here then is his conclusion:

> I think it impossible that a mind, free from all preconception, can reflect upon the extreme diversity of the phenomena which thus converge to the same result, without experiencing a very strong impression, and I think it will henceforth be difficult to defend by rational arguments a hostile attitude to molecular hypotheses, which, one after another, carry conviction, and to which at least as much confidence will be accorded as to the principles of energetics (1910a, 91; cf. also b, 281).

In light of this, I very much doubt that Perrin's attitude towards the molecular theory was the one suggested by van Fraassen. There is nothing objectionable *per se* in van Fraassen's (2008, 112) claim that "*experimentation is the continuation of*

---

[5] A distinct part of Perrin's work on the molecular explanation of Brownian motion was related to his attempt to verify experimentally Einstein's (1905) theory of diffusion. The relation of Perrin's work to Einstein's is discussed in my (2011). Perrin offers a very detail discussion of Einstein's theory and his own experimental verification of it in his 1911 Solvay conference paper (cf. 1912, 189–216).

theory construction by other means." But experimentation is not just that! It is also (surprise!) a vehicle for testing theory and rendering it probable. In the discussion that followed Perrin's address to the French Philosophical Society, Perrin (1910b, 300) made a related point very forcefully:

> In reality this antagonism [between atomism and its opponents] translates two tendencies that oppose each other rather deeply: one that encourages us to make hypotheses in order to go forward, and another which warns off all suppositions that cannot inspire in us an immediate experiment which is immediately feasible. The energeticists classify the experiments and generalise them algebraically; the atomists look for ways to penetrate the mechanism, to go beyond it, and for this they imagine molecules ... briefly put, they set themselves the problems that the energeticists consider to be superfluous and purely apparent. Let Brownian movement be h. Our experiments reveal a function of certain measurable quantities: $h = f(R, R')$. And also another function is given to us: $h = \varphi(A, A')$ (energy). If you fear the apparent problems, it suffices not to assume the existence of h and to set $f = \varphi$. The atomists, by contrast, try to guess something behind these functions, that is to say, to specify h.

A somewhat stronger point was made by Louis Couturat (Perrin 1910b, 293), who was present in the meeting alongside many others major French philosophers of the time. After claiming that one could see Perrin's achievements as aiming to ground on controllable and measurable facts a bunch of fictions, he offered the following riposte on Perrin's behalf:

> The fact that my [Perrin's] hypotheses are adapted to reality, and in so many different ways, this is what one calls an experimental verification; this is for us physicists the criterion of truth. Call that language whatever you want, always it is our language framework that 'sticks' to the facts.

7.5 The View from Above

What would be a reasonable way to spell out the logical structure of Perrin's argument? Recall his claim that he was after a *crucial experiment* for the reality of atoms. Of course, there are no crucial experiments in the strict sense of the expression, viz., in the sense of disproving a hypothesis or of proving a hypothesis. But as Poincaré has put it, an experiment can condemn a hypothesis, even if it does not — strictly speaking — falsify it. Perrin's argument was precisely meant to condemn the denial of the atomic hypothesis, viz., that matter is continuous.

The way I think Perrin's argument should be reconstructed is as follows. With the reasoning sketched in the previous section, Perrin has made available two important probabilities, viz.

$$\text{Prob}(n{=}N/AH) = \text{very high}$$
$$\text{Prob}(n{=}N/\text{-}AH) = \text{very low}$$

That is, the probability that the number n of molecules in a gram-molecule of a gas (including an emulsion, which does behave as a gas) is equal to the Avogadro number N given the atomic hypothesis is very high, while the probability that the number of molecules n is equal to the Avogadro number N given the denial of the atomic hypothesis is very low.

To see why Prob(n=N/−AH)=very low, recall Perrin's point (stressed in his address to the French Philosophical Society) that on the negation of atomic hypothesis, the predicted value of N could be *anywhere* between zero and infinity, which means that the probability that the predicted value of n would be *equal* to N would be zero. This does not mean that the negation of the atomic hypothesis implied that matter is discontinuous, with n≠N! Rather the negation of the atomic hypothesis (in the specific form advocated by Perrin, where n=N) is consistent with *any* value of N from 0 to infinity.

These two likelihoods (in the technical sense of the term) can be used to specify the Bayes factor f.

$$f = \text{prob}(n=N/-AH)/\text{prob}(n=N/AH)$$

Bayes's theorem states

$$\text{prob}(AH/n=N) = \text{prob}(n=N/AH)\,\text{prob}(AH)/\text{prob}(n=N)$$

where:

$$\text{prob}(n=N) = \text{prob}(n=N/AH)\,\text{prob}(AH) + \text{prob}(n=N/-AH)\,\text{prob}(-AH)$$

Using the Bayes factor, Bayses's theorem becomes this:

$$\text{prob}(AH/n=N) = \text{prob}(AH)/(\text{prob}(AH) + f\text{prob}(-AH)).$$

Perrin's argument then can be put thus:

1. f is very small.
2. N=n is the case.
3. prob(AH) is not very low.

Therefore, prob(AH/n=N) is very high.

Premise 1 (that f is very small) is established by the body of Perrin's demonstration, which shows that given the denial of the atomic hypothesis, it is extremely unlikely that the Avogadro number has the specific value it does (that is, that the number n of molecules in a gram molecule of a gas is equal to Avogadro's number N). Premise 2 is established by a series of experiments involving different methods and different domains. Premise 3 is crucial, since it is required for the probabilistic validity of Perrin's argument. It specifies the prior probability of the atomic hypothesis and without the prior probability the argument noted above would commit the base-rate fallacy. Perrin's preparatory eliminative work had aimed to show that, by eliminating several alternative potential explanations of Brownian movement, the atomic hypothesis had gained at least some initial plausibility which was reflected in its having some prior probability of being true.

This kind of reconstruction would (and does) explain Perrin's own *confidence* that the atomic hypothesis has been "established;" that he has offered "a decisive proof" of it (1916, 104). Admittedly, some reliance on the prior probability prob(AH) is inevitable and the usual philosophical dialogue would kick off: How are the priors fixed? Are they *really* objective? If not, is the case for the reality of atoms really strong?

It is certainly arguable that prior probabilities express reasonable degrees of belief, which supervene on certain causal and explanatory qualities of a given hypothesis. So, prior probabilities need not be purely subjective or idiosyncratic degrees of belief — though reasonableness need not be determined in any algorithmic way.[6] In any event, the case in hand is quite peculiar for the following reason. It presented to anyone involved (and mainly to working scientists) a case in which the posterior probability of the atomic hypothesis becomes (almost) unity — given, of course, that it is assigned a *non-zero* prior probability, which it seems everybody but Duhem did.[7] This case might well be exceptional, but its role in the establishment and wide acceptance of the atomic conception of matter can hardly be underestimated.

There is another feature of Perrin's strategy that needs to be highlighted. In his (1916, 105) he claims:

> The objective reality of the molecules therefore becomes hard to deny. At the same time, molecular movement has not been made visible. The Brownian movement is a faithful reflection of it, or, better, it is a molecular movement in itself, in the same sense that the infra-red is still light.

Perrin's point here is precisely that size does not matter, but causal role does! Like Pasteur before him, Perrin did place the molecules firmly within the laboratory, grounding their causal role and offering experimental means for their detection and the specification of their properties — even though, the molecules did not become

---

[6] The whole issue is delicate, of course. But I think the following dilemma should be resisted: *either* prior probabilities should be fixed in a fully objective and logical manner (God-given? based on purely logical or synthetic a priori principles like the Principle of Indifference?) or else they are purely subjective and idiosyncratic and therefore useless in the defence of the rationality of the belief in theories. A priori probabilities can be whimsical, but they need not be. They can be based on judgements of plausibility, on explanatory considerations prior to the collection of fresh evidence and other such factors, which — though not algorithmic — are quite objective in that their employment does and should command rational agreement (As Perrin's case nicely illustrates).

[7] Is there any reason to take the broadly Bayesian reconstruction of Perrin's argument as being Perrin's own? There is some interesting circumstantial evidence, coming mostly from the fact that Émile Borel — who was Perrin's close friend and colleague — was an expert on probability theory and had actually made explicit reference to Bayes's theorem in his (1914). Borel (1914, 99) explicitly associated Bayes's theorem with the case of finding the probability of the causes (given their effects) and stressed that there is need to specify the a priori probability of the cause, though he admitted there was uncertainty as to how a priori probabilities were estimated. Borel did make many references to Perrin's statistical methods in his (1914).

visible. This is of great significance because it becomes clear that Perrin's argument should be compelling for anyone who does not take it that strict naked-eye observability is a necessary condition for accepting the reality of an entity. It should also be compelling for anyone who thinks that continuity of causal role is a criterion for accepting the reality of an entity — irrespective of whether some instances of this entity are observable, while others are not. Recall Perrin claim that the movement of the Brownian particles was a "faithful reflection" of the molecular movement, since the Brownian particles *were* large molecules. Perhaps, then it becomes clear why Perrin's argument could persuade almost everyone but Duhem, who took a very hard line on observability and denied the call for explanation-by-postulation.[8]

7.6   van Fraassen's Perrin Evaluated

van Fraassen does not really discuss Perrin's theoretical model in any detail, but he (2009, 20) does mention Perrin's claim that the Brownian particles behave as large molecules and hence obey the laws of gases. He adds:

> Perrin argues for its plausibility, but in terms that clearly appreciate the postulational status of this step in his reasoning. (…) On this basis, the results of measurements made on collections of particles in Brownian motion give direct information about the molecular motions in the fluid, always of course within the kinetic theory model of this situation. But that is just what was needed for empirical grounding of those remaining theoretical parameters.

As noted already, there need not be a tension between the need to ground empirically a theory and its being taken to be a plausible, or even a probable, theory. Perrin did try to ground empirically the atomic theory, but he did not try to do *just this* — at least in the way van Fraassen reads the claim of empirical grounding. It is perfectly consistent to try to ground empirically some theory and to claim that this theory is true, or by and large true (or at least that is highly confirmed by the measurements that ground it empirically). It is precisely this kind of stance that should be attributed to Perrin. Hence, empirical grounding — which turns out to be necessary for the enhanced testability of the theory — is a means to a broader end, viz., the confirmability of the theory. In Perrin's hand, the atomic-theory-based account of the Brownian motion did not just end up being confirmable, but was actually confirmed by a striking prediction of Avogadro's number.

In light of van Fraassen's overall stance, it might be tempting to think that it is enough to say of Perrin's strategy that it was aiming to show that the molecular hypothesis was empirically adequate. Or it might be that it was just aiming to lay to rest "the idea that it might be good for physics to opt for a different way of modelling nature, one that rivalled atomic theories of matter" (2009, 23). But here again, this kind of reading — especially in the latter form — is fully consistent with Perrin

---

[8] Some more general lessons for scientific realism that can be drawn from this case are discussed in my (2011).

aiming for more — as, in fact, he did. Actually, more can be said. We saw Perrin striving for an articulation of the theoretical mechanism (model) by means of which the all-important exponential law (1) was achieved. He could have started with the exponential law itself, without seeking to explain it. Striving for an explanation/ grounding of this law simply does not make sense unless Perrin was aiming establish the causes of the Brownian motion.

There is a seemingly unexpected twist in Perrin's story that could not have escaped van Fraassen's acute attention. Didn't Perrin end up his (1910a) with the strange claim that the reference to molecules was dispensable? Indeed, he (1910a, 91) stressed the following:

> Lastly, although with the existence of molecules or atoms the various realities of number, mass, or charge, of which we have been able to fix the magnitude, obtrude themselves forcibly, it is manifest that we ought always to be in a position to express all the visible realities without making any appeal to elements still invisible. But it is very easy to show how this may be done for all the phenomena referred to in the course of this Memoir.

And then he proceeded to show how the very reference to Avogadro's number can be eliminated. Consider any two laws in which N features as a constant (e.g., Einstein's diffusion equation and the law of the distribution of radiation) and take their pure functional form. "The one," Perrin says,

> expresses this constant [Avogadro's number] in terms of certain variables, a, a', a'', …,

$$N = f[a,a',a'',...];$$

> the other expresses it in terms of other variables b, b', b'', …,

$$N = g[b,b',b'',...].$$

> Equating these two expressions we have a relation

$$f[a,a',a'',...] \equiv g[b,b',b'',...],$$

> where only evident realities enter, and which expresses a profound connection between two phenomena at first sight completely independent, such as the transmutation of radium and the Brownian movement (1910a, 91–92).

This way to proceed might well suggest that, in the end, Perrin wanted to show that the molecular hypothesis is eliminable: a scaffolding that may well be removed after connections between empirical phenomena have been established. van Fraassen does not quite put it like this and he warns us not to read the above passage in a philosophically loaded way. He adds:

> I do not offer this [passage] as a case of an apparent scientific realist contributing grist for the empiricist's mill! Rather, this passage is important because of how it illustrates the factors of *Determinability* and *Concordance* in empirical grounding.

This peace-offering however is unnecessary. Note, for one, that the "evident realities" which enter into the functional relations thus established are not merely observable magnitudes or properties of observable entities. On the contrary, it is evident that they are not. For instance, the diameter of the Brownian particles or the wave-length of emitted light are not observable. They are, however, determinate and measurable and this is what Perrin insisted one. More importantly, however, Perrin did not take it that the possibility of eliminating the constant N implied that molecules could be dispensed with. In (1910a, 92) he noted that the discovery of functional relations such as the above — which could not have been established without the atomic hypothesis — mark "the point where the underlying reality of molecules becomes part of our scientific consciousness."

And if this left any doubt to his reader about his commitments, in two subsequent publications in which he also presented *verbatim* the same idea of establishing functional relations among "evident realities" he added:

> But, under the pretence of rigour, we will not make the mistake to throw thus out of our equations the elementary magnitudes that allowed ourselves to obtain them. This would not be to remove a scaffolding that has become useless to the finished structure; it would be to mask the pillars that that have made its skeleton and beauty (1912, 250).

And in his *Les Atomes*, he put the point in a similarly graphical way:

> But, under the pretence of rigour, we will not make the mistake to avoid the operation of the molecular elements in the enunciated laws that we would not have obtained without their assistance. This would not be to uproot a useless stake from a thriving plant; it would be to cut the roots that nourish it and make it grow.[9]

Indeed, it seems it does not make good sense to read Perrin's claim about the role of relations of the form $f[a, a', a'', \ldots] \equiv g[b, b', b'', \ldots]$, as van Fraassen does, as illustrating the factors of concordance and determinability. The various ways to specify Avogadro's number lack any kind of concordance unless they are taken to determine *Avogadro's number*; what were concordant were precisely the values of N, as they were determined in various ways. Given that the access to the molecules is only indirect — and given Perrin's insistence that their magnitudes should be determined and measurable — it was important to be shown that these magnitudes are essentially invariant irrespective of the observable phenomenon that leads to their calculation. The invariance of Avogadro's number was the key to proving the invariance of the molecular properties.[10]

Insofar as van Fraassen intends to hold on the general view that disbelief in a theoretical hypothesis is *always* a reasonable option, Perrin's case — looked at both from within and from above — shows that it is not. There are cases in which asserting the reality of certain entities is the only reasonable option.

---

[9] My translation from p. 284 of the French edition of *Les Atomes* (Flammarion, 1991). The rendering of this passage in the English translation of the book (1913, 207) is mistaken.

[10] A version of this point is made by Louis de Broglie (1945, 11).

7.7 Brownian Movement Was *not* a Public Hallucination

van Fraassen (2009) takes it that scientific instruments are not "windows on the invisible world" but rather "engines of creation" of new observable phenomena that theories have to save. In the case of microscopes, van Fraassen makes the rather astonishing move to consider the phenomena thus created to be "public hallucinations." Far from giving an image of some unobservable-by-the-naked-eye entities, the image seen in a microscope is just an image. Not an image of anything, but a public hallucination. The rainbow, van Fraassen says, is a public hallucination — there is no *bona fide* object that is a rainbow. It lacks certain invariances; it has no spatio-temporal position etc. But, he says, the images seen under a microscope too are public hallucinations. As he put it, "Nature creates public hallucinations" (2009, 103) and microscopes "imitate the ability of nature to create public hallucinations" (2009, 104). To be more precise, public hallucinations are

> a whole gallery of images which are not things, but are also not purely subjective, because they can be captured on photographs: reflections in the water, mirror images, mirages, rainbows (2009, 105).

Some of them, van Fraassen adds, are "copy qualified" in the sense that they lend themselves to being interpreted as images of real things. Microscope images, unlike rainbows and mirages, are copy-qualified: it makes sense to ask of them whether they represent something real or not.

His view, however, is that

> the microscope *need not* be thought of as a window, but is *most certainly* an engine creating new optical phenomena. It is accurate to say of what we see in the microscope that we are "seeing an image" (like "seeing a reflection," "seeing a rainbow"), and that the image could be *either* a copy of a real thing not visible to the naked eye or a mere public hallucination. I suggest that it is moreover accurate and in fact more illuminating to keep neutrality in this respect and just think of the images themselves as a public hallucination (2009, 109).

Why is keeping neutrality, one may wonder, more accurate and more illuminating? To show that it is not, let us look once more at the case of Brownian motion: the random and incessant motion of microscopic particles suspended in a liquid was observed through a microscope. Think of it as an image on the lens of the microscope. Let us state some of its properties.

First, the image co-varies with something else, viz., the liquid drop which is observed: if the liquid drop is removed, so is the image. So there is a correlation between the image and something else. Actually, there should be no doubt that this something else — call it X — (the liquid with the suspended microscopic particles in our case) *produces* (or at least essentially contributes to the production of) this image. Even if X were not the total cause of the image, it would be a substantial part of the cause since by removing X, the image is removed. It might be that when X is present, something else Y is present too — e.g., a distorting effect of the lens. But even then, Y could not produce the effect on its own. Let's call this *regularity*.

Second, the image is definite enough to be distinguished from other images that have *prima facie* similar causes — e.g., the image of liquid drops with no particles suspended in them. Let's call this *definiteness*.

Third, the image presents a certain temporal fixity. It can be present and be observed on slides that had been conserved for decades under all kinds of external conditions (cf. Nye 1972, 24). Let's call this *resilience*.

Fourth, the image displays certain important invariances. For instance, Robert Brown himself noted that the random movement occurs when pollen of various plants were used. In fact, between 1830 and 1870, physicists and biologists used a great variety of organic and inorganic particles: sulphur, mastic, cinnabar, pulverised coal, India ink and gamboge and they suspended in a variety of fluids (see Nye 1972, 23). Let's call this *invariance*.

Fifth, the image does not go away whenever certain factors are involved in the preparation of the emulsion: sunlight and darkness, electricity and magnetism, temperature variations etc. (cf. Nye, op.cit.) Let's call this *robustness*.

Sixth, the image is manipulable — by manipulating its causes. Perrin was an actual master of this. He used various materials (e.g., gamboge and mastic); he prepared the granules in various meticulous ways; he used various methods to avoid sources of error etc.[11] Let's call this *manipulability*.

I very much doubt that these properties can be had by the rainbow. Or by anything which cannot reasonably be taken to be "copy qualified" image. But let's not argue about this directly. Let's take an indirect route. It was exactly the possession of properties like these that rendered necessary (and desirable) an explanation of the Brownian images. The explanation could proceed at two levels — one intrinsic, the other extrinsic. At the intrinsic level, it would have to be an explanation of the image in terms of the properties of the causes of the image — that is, of the properties of the liquid and of the suspended in it particles. It would require thinking of the image as an image of something — of what is going on *within* the liquid — and would proceed by eliminating various hypotheses as to what is going on within the liquid by eliminating, in effect, various competing images that one would have expected to see in the microscope had the alternative hypotheses of the origin of the image been true. This is more or less the actual course of events, until and during Perrin's experimental work on the Brownian motion. Its very possibility is predicated on taking the image (of Brownian motion) to be the image of whatever it is *within* the liquid that causes it; to be, an image, as Perrin put it, of the internal agitation of the fluid (cf. 1910a, 5). The image did not render visible the molecular movement, but as Perrin himself put it, it was nonetheless a "faithful reflection" of it since the Brownian particles, of which there were copy-qualified images, were large molecules.

---

[11] Perrin presents in painstaking detail the various ways in which he manipulated the emulsions that he studied and his various attempts to establish concordances between the values of the properties of the Brownian particles (cf. 1910a, §§15–22). At one point he described how he had to wait for 2 or 3 days for various protozoa to die that had developed in an emulsion which had not been rendered aseptic. The bacteria "fell inert to the bottom of the preparation" (1910a, b, 41). For an illuminating discussion of the use of the ultramicroscope from Perrin, cf. Bigg (2008).

The other level at which the explanation of the Brownian images could proceed, as I already said, would be extrinsic. Note that by that I do *not* mean the search for external causes — that is causes of the observed random motion of the Brownian particles that operate outside the liquid, e.g., the road traffic, the effects of which on the microscope a few experimenters tried to shield. This is absolutely fine and intrinsic *in the above sense*. It is an attempt to exclude alternative hypotheses that would have led to alternative images, had they been true. What I mean by "extrinsic explanation" of the image is an explanation of why *this* particular image arises as opposed to anything else. And this would be an explanation of why observers like us see an image like that when they place their eyes in contact with a microscope and a certain liquid is put on the film. This course of action would be absolutely natural, had there been thought that the Brownian image was a public hallucination. Then, the course of the explanation would be very much like the course of the explanation of why we see the rainbow while there is no such thing as the rainbow (or similarly why we see the blue sky though there is no such thing as the blue sky). Differently put, the intrinsic explanation would aim to answer the question of why there is an image like this — and would proceed by examining what it is an image of. The extrinsic explanation would proceed by aiming to answer the question of why we *see* an image like this — and would proceed by examining the causes of our *seeing* it. The intrinsic explanation — the one that was actually pursued — required thinking of the image as a copy-qualified image, while the extrinsic explanation would require thinking of the image as a public hallucination to be explained away *qua* an image of anything.

In the case of Brownian motion, van Fraassen's recommended neutrality is neither illuminating nor accurate. It certainly does not tally with viewing the history of the work on the Brownian motion *from within*.

It might be ironic that when Emile Meyerson (1912 [1930], 90) discussed the shift of opinion in favour of the atomic conception of matter, he noted the following:

> (A)t first sight one is almost tempted to ask whether these investigators [of the Brownian movement] have not been victims of an illusion in this case, if they have not succumbed to an unconscious trick of their own minds.

But he immediately added that looking at the results Perrin had produced entirely frees is from these doubts.

7.8   Merging the Two Views

There is no theory-free perspective on reality. But this does not mean that there is no way to form a reasonable belief about what reality is like. Theories are apt for confirmation and well-confirmed theories do offer good reasons (based on the link between theory and evidence) to think of reality as being in a certain way. van Fraassen is right when he stresses that a God-like view of nature is impossible, but wrong when he takes it that, *because of this*, the deep-structure of reality is impenetrable. Denying the impossible should not make us blind to the possible. When all goes well, theory and experiment — or measurement practices — develop hand in hand. Theory is under constant pressure to render its theoretical parameters determinate and measurable. Measurement both makes theory develop and tests it. When viewed *from within*, theories develop by being in constant interaction with experience. When viewed *from above*, theories are assessed on the basis of experience. The stereoscopic view we have been looking for aims to combine the process and the product. Perrin's case shows that this stereoscopic view need not leave us in the dark as to what the texture of reality is. Indeed, van Fraassen's narrative of Perrin's achievement refuses — ultimately — to view the theory from above; that is, to unravel the general reasoning pattern which made Perrin's achievements so decisive in turning the balance in favour of the atomic conception of matter.

I just want to repeat it: there are cases in which asserting the reality of certain entities is the only reasonable option. This is Perrin's case. Might that be too strong a claim to make? Historically, it has been brought out. Duhem's denial till the bitter end was based on philosophical dogmatism — in essence, in assigning a zero prior probability to the atomic conception of matter, based on the claim that it's primarily an explanatory hypothesis and such hypotheses fall outside science. But even if we leave the actual history out of the picture — as we should *not* — the broader philosophical point is this. What Meyerson aptly called "impartial observers" are precisely those scientists or philosophers who are epistemically open: they can change their minds when decisive evidence in favour of (or against) a hypothesis becomes available — evidence that meets ordinary scientific criteria of relevance and strength. I find it hard to believe that an impartial observer of Perrin's achievements could reasonably resist Perrin's conclusions.

References

Bigg, C. (2008). Evident atoms: Visuality in Jean Perrin's Brownian motion research. *Studies in History and Philosophy of Science, 39*, 312–322.

Borel, E. (1914). *Le Hasard* (2nd ed., 1920). Paris: Librairie Félix Arcan.

Brush, S. (1968). A history of random processes. *Archive for History of Exact Sciences, 51*, 1–36.

de Broglie, L. (1945). *La Réalité Des Molécules et L'Œuvre de Jean Perrin* (Académie des Sciences). Paris: Gauthiers-Villars.

Einstein, A. (1905 [1956]). On the movement of small particles suspended in a stationary liquid demanded by the molecular-kinetic theory of heat. In R. Furth (Ed.), *Investigation on the theory of the Brownian movement* (pp. 1–18). New York: Dover Publications.

Glymour, C. (1975). Relevant evidence. *The Journal of Philosophy, 72*, 403–426.

Gouy, L. (1895). Le Mouvement Brownien et les Mouvement Moléculaires. *Revue Générale des Sciences Pures et Appliquées, 6*, 1–7.

Meyerson, E. (1912 [1930]). *Identity and reality* (K. Loewenberg, Trans.). New York: Macmillan.

Nye, M. J. (1972). *Molecular reality: A perspective on the scientific work of Jean Perrin*. London: MacDonald.

Nye, M. J. (1976). The nineteenth-century atomic debates and the dilemma of an indifferent hypothesis. *Studies in History and Philosophy of Science, 7*, 245–268.

Ostwald, W. (1896). The failure of scientific materialism. *Popular Monthly, 98*, 589–601.

Ostwald, W. (1907). The modern theory of energetics. *The Monist, 17*, 481–515.

Ostwald, W. (1912). *Outlines of general chemistry* (3rd ed., W. W. Taylor, Trans.) London: MacMillan.

Perrin, J. (1901). Les hypothèses moléculaires. *Revue Scientifique, 15*, 449–461.

Perrin, J. (1903). *Traité de Chimie Physique: Les Principes*. Paris: Gauthier-Villars.

Perrin, J. (1910a). *Brownian movement and molecular reality*. (F. Soddy, Trans.). London: Taylor and Francis.

Perrin, J. (1910b). Le Mouvement Brownien. *Bulletin de la Société Française de Philosophie, 10*(4), 265–302. Séances du 27 Janvier et du 3 Mars 1910. Paris: Vrin.

Perrin, J. (1911). La Réalité des Molécules. *Revue Scientifique, 25*, 774–784.

Perrin, J. (1912). Les Preuves de la Réalité Moléculaire (Etudes Spécial des Emulsions). In P. Langevin & L. de Broglie (Eds.), *La Théorie Du Rayonnement et les Quanta* (pp. 153–253). Paris: Gauthier-Villars.

Perrin, J. (1916). *Atoms*. (D. L. Hammick, Trans.). London: Constable and Company Ltd.

Poincaré, H. (1900). Les Relations Entre la Physique Expérimentale et la Physique Mathématique. In C.-E. Guillaume & H. Poincaré (Eds.), *Rapports Présentés au Congrés International de Physique de 1900* (Vol. 1, pp. 1–29). Paris: Gauthier-Villars.

Poincaré, H. (1906). The principles of mathematical physics. In H. J. Roberts (Ed.), *St Lewis International Congress of Arts and Science* (pp. 604–622). London: University Alliance.

Poincaré, H. (1913). *Mathematics and science: Last essays*. New York: Dover.

Post, H. R. (1968). Atomism 1900 I & II. *Physics Education, 3*, 225–232 and 307–312.

Psillos, S. (2011). Moving molecules above the scientific horizon: On Perrin's case for realism. *Journal for General Philosophy of Science, 42*, 339–363.

Thomson, W. (1870). On the size of atoms. *Nature, 1*, 551–553.

van Fraassen, B. (2008). *Scientific representation: Paradoxes of perspective*. Oxford: Clarendon Press.

van Fraassen, B. (2009). The perils of Perrin, in the hands of philosophers. *Philosophical Studies, 143*, 5–24.

Weyl, H. (1927 [1963]). *Philosophy of mathematics and natural science*. New York: Atheneum.

Part IV
Scientific Explanation and Epistemic Values Judgments

Chapter 8
Explanation as a Pragmatic Virtue: Bas van Fraassen's Model

Margarita Santana

Abstract This work provides an analysis of van Fraassen's model of explanation in the theoretical framework of the scientific explanation models. The objective is, first, to see his contribution to this framework and, secondly, what objections or criticism he is capable of. The analysis focuses, in this sense, in exposing the characterization that provides the explanation as a pragmatic virtue to determine if, indeed, the model proposed by van Fraassen, that is the first model to take elements as actors, contexts, and audiences into consideration, can be considered a pragmatic model of explanation. The aim is also to show that the theorisation of explanation incorporates underlying theorisations which determine the way in which explanation is conceptualised within each proposed model.

Keywords Scientific explanation • Models of explanation • Explanation as a pragmatic virtue • Contextual factors • The pragmatics of explanation

8.1 Theoretical Context

Scientific explanation is an element of the basic repertoire of what is now known as classical philosophy of science. It is a key notion which has articulated philosophical thinking regarding science almost from its very beginning,[1] and which, to my mind, contains one particular quality of inestimable value: the fact that its theorisation incorporates and deploys other thematic nuclei which come into play and which

---

[1] One simply has to remember, for example, the Aristotelian reflections regarding science — demonstration — with its cognoscitive and explanatory dimensions. See Aristotle (1988) and also Vega (1990).

M. Santana (✉)
Faculty of Philosophy, University of La Laguna, Guajara Campus, w/n,
38296 La Laguna, Tenerife, Spain
e-mail: msantana@ull.es

W.J. Gonzalez (ed.), *Bas van Fraassen's Approach to Representation and Models in Science*, Synthese Library 368, DOI 10.1007/978-94-007-7838-2_8,
© Springer Science+Business Media Dordrecht 2014

are referred to either implicitly or explicitly. The different models or theories which are proposed as elucidations of this concept, i.e., as "explanations" of the concept of *explanation*, are models which, at the same time, provide specific canons of scientific explanation, and the analysis of these different elucidations reveals (and this is the principal idea being developed here) that the theorisation of explanation incorporates underlying theorisations which determine the way in which explanation is conceptualised within each proposed model. In a previous chapter, I charted the different ways in which this theme has been dealt with since its founding moment with Hempel and Oppenheim.

Hempel's model of explanation constitutes an undeniable point of reference as a hegemonic model associated with a certain conception of an (also hegemonic) philosophy of science up until around the 1960s. Indeed, the different models which appeared after this period (that of Salmon, Kitcher, Nancy Cartwright, van Fraassen and Achinstein, to cite but a few) all refer, to a greater or lesser extent and in addition to their own contributions, to this common Hempelian point of reference. In this sense, in the trajectory of this concept, different movements intertwine and overlap, spilling over the boundaries of the models themselves. Thus, based on my reconstruction of this evolution, a portrait or description can be gleaned in which the different emphases vary, from the centrality of the logical structure of explanations to the importance of the causal structure of the world, to which explanations must refer and into which they must fit, as well as the priority bestowed on the pragmatic aspects in comparison to others, such as the logical or semantic ones.

This difference in emphasis, which itself is a symptom of deeper-rooted differences regarding the underlying theorisations mentioned above, points nevertheless to a single trend: the most notable consequence of the increasing importance of pragmatic approaches is the need to consider the inescapable significance of actors, contexts and audiences, and to understand the difference between process and product as a specific difference that enables the analysis of science (hitherto limited to its dimension of end result, consolidated knowledge and product) to be expanded, enriched and complemented. The model of explanation proposed by van Fraassen is the first model to take all these elements into consideration.

8.2   The Concept of "Scientific Explanation"

Science has an explanatory dimension which, in general terms, we can sum up as follows: it "accounts for" certain events or phenomena, it "explains" them. This dimension is not exclusive to science, although in the sense that it is in this field in which our knowledge of the world acquires its maximum expression, the "scientific explanation" of the world can be seen, perhaps, as the "quintessential" method of explanation. Also, and again in general terms, science is not only concerned with *what*, it also focuses, specifically on *why*[2]; and in responding to the *why* of things, it

---

[2] This first statement may seem a rather circular way of explaining the reasons why scientific explanation is the quintessential means of explanation, because it boils down to the classical

uses its theories and accounts of the world, and by doing so, it provides a genuine and deep-rooted understanding of the world. In other words, it provides information about certain things.

This first characterisation is by no means exhaustive or without its problems, but it does enable us to understand that, in the context of philosophy of science, explanation has long constituted a thematic nucleus which presents a series of peculiarities or specificities which have enabled it to gain a level of importance almost impossible to dispute. The reasons for this importance are mainly related to two issues: firstly, the consideration of explanation as one of the principal objectives or goals of science[3]; and secondly, the idea that theorisation regarding explanation incorporates and deploys all the scientific products currently available at any given time: "scientific explanation constitutes the alpha, in relation to the objectives of science, and the omega, as regards the conceptualisation which encompasses and puts all other conceptualisations into play." (Estany 1993, 229.) In other words, scientific explanation incorporates other thematic nuclei (concepts, laws, theories, models, types, etc.), which enter into play and which are referred to either implicitly or explicitly.

Regardless of the general agreement regarding the reasons given for the importance of the concept of "explanation," and even regarding this importance itself, there are a number of different analyses (theories or models) of this concept which, from the perspective of general meta-scientific research, aim to provide an elucidation; i.e., they attempt to give an "explanation" of the concept of *explanation*,[4] while at the same time offering specific canons of scientific explanation.

Although we can distinguish between these two tasks for the purposes of analysis, they are in fact closely related, because the determination or characterisation of whatever counts as an explanation already encompasses the elements in virtue of which the whole process is carried out. The differences existing between different models, even those with the same stated objective, may be due, for example, to the fact that they adopt different perspectives when characterising explanation: in accordance with its logical structure, the type of statements intervening in it, different sciences, etc. Nevertheless, it is perhaps possible to establish the existence of a common conceptual nucleus which links them all together. This conceptual nucleus would be as follows: in any explanation we can identify three elements: the *explanandum*, i.e., that which requires an explanation; the *explanans*, i.e., that which provides the explanation of the *explanandum*; and the explanatory relationship, i.e., the relationship which, by existing between the *explanans* and the *explanandum*, enables us to consider that the former explains the latter.

---

distinction between description and explanation: science not only describes, it also explains. However, this must be understood in relation to the following statement regarding how it responds to the question "why?."

[3] As we will see later on, this view of explanation, although perhaps the object of almost unanimous agreement at first, has now been called into question by different approaches.

[4] In this sense, see Díez and Moulines (1997, 220), who distinguish between "explanation" as an explanation of any meaning, and "explication" as the elucidation of a concept.

Given this configuration, we could state that any analysis of the concept of explanation should provide a precise characterisation of these three elements. Thus, one way of approaching the study of the different models is to examine the way in which they conceive and analyse said elements in their specific contexts. Also, the differences between these models may be interpreted and explained in accordance with the different ways in which these elements are conceived and analysed (Díez and Moulines 1997, 224).

While the different models or theories of explanation aim to elucidate the concept of *explanation*, this paper aims to analyse a specific model of explanation, namely that proposed by van Fraassen. Thus, I shall divide the analysis into two levels or phases: the first will take as its starting point the aforementioned conceptual nucleus, and the aim is to present an exposition or description of the model, focusing exclusively, and as far as possible, on the extent to which it satisfies this nucleus; and the second will aim to present some of the criticism which has been levelled at the model, in order to show that said criticism can be interpreted as the expression of differences in the analysis of the nucleus elements, and that, in fact, it is the result of factors which underlie or are implicit in each model which, although not related *prima facie* with explanation as such, nevertheless determine the way in which this notion is conceived. In other words, the aim is to show that theorisation on explanation incorporates underlying theorisations which determine the way in which the notion is conceptualised in each proposed model.

We can start by highlighting the fact that although explanation has an epistemological dimension, its consideration in these terms fails to explore its full meaning, because it also includes always conceptions and assumptions regarding (or reaching beyond) purely epistemic aspects. Moreover, we could say that its epistemological dimension can only be defined with any precision thanks to these other assumptions. By "other assumptions" I refer mainly to two questions: the concept held of scientific knowledge, which *grosso modo* includes the conception of scientific theories or that of scientific activity itself, in the sense of theoretical activity and intervening activity, the precedence of one over the other or the symmetrical consideration of both, etc.; and the way in which the relationship between knowledge and the world is understood, which may in turn determine the way in which we understand the explanatory relationship in the context of the characterisation of that which is distinctive about explanatory knowledge,[5] and which reveals the ontological commitments one is willing to undertake (or not) when accounting for scientific knowledge.

Thus, alongside the specification or determination of nature and the role played by explanation in science, there is also the determination or specification of the suppositions made (or not) by the scientific explanation of the world regarding this. In this sense, our basic idea, which I propose as an analytical tool or heuristic instrument, is that the different conceptions of explanation reflect the confluence of the

---

[5] At least in those cases in which the distinction between description and explanation is assumed and defended, and in which a description is given of the traits which characterise the latter in relation to the former.

aforementioned factors, and that these in turn determine, within each model, the way in which said conceptions are configured.

8.3 Explanation from a Pragmatic Perspective

The Scientific Image (van Fraassen 1980 [1996]) can (among other possibilities) be interpreted as an original contribution to the realism-antirealism debate, or, more specifically, as a way of cancelling or transcending it. I. Hacking (1983, 166–167) had already stated that realism is a problem, because we have conceived alternative styles of representation; in other words, the problem arises because we have alternative systems of representation. For the same reason, if only one type of representation were to exist, antirealism would make no sense. According to the author, both positions navigate the field of representation, trying to find in its nature something which enables them to dominate the other, "but there is nothing else there." B. van Fraassen also bases his argument on the possibility and existence of alternative representations, of different theories for the same set of phenomena, and believes that the only thing one can demand of them is empirical adequacy. In relation to the other questions, the best option is to suspend judgement and to adopt an agnostic attitude, since they only add a totally unnecessary metaphysical dimension to thinking about science. In this sense, van Fraassen provides a model of explanation in accordance with this vision, which aspires to transcend the different problems associated with the approaches to explanation proposed to date.

8.3.1 Explanation as a Pragmatic Virtue: van Fraassen's Model

van Fraassen initiates his approach to the theme of explanation by applying to scientific theories the division that Ch. Morris postulated for language. According to this division, three different levels can be discerned: the syntactic level, the semantic level and the pragmatic level. Similarly, in theories we find a threefold division of relations and properties, including the purely internal or logical ones (equivalent to the syntactical level and exemplified by axiomatisability, consistency and various types of completeness), semantic ones (i.e., those concerned with the relationship between the theory and the world[6]; principal properties here would be truth and empirical adequacy) and finally, pragmatic ones, in relation to which he introduces an important clarification. The author concedes that scientific theories may be stated in a language which is independent of context, but scientific activity itself, understood as a framework encompassing more elements than just pure theory, includes

---

[6] Or, more specifically, as the author himself points out: "(…) the facts about which it is a theory (van Fraassen 1980, 90)."

two aspects in which the language used is radically context-dependent[7]: the language of theory evaluation, and specifically the term "explains," and the language of the use of theories to explain phenomena.

In this sense, when one defends, assesses or puts forward reasons for preferring one theory over another, the only virtues to consider in relation to the world would be consistency and empirical adequacy and scope; all other virtues, including explanatory capacity, would be pragmatic virtues, i.e., virtues relative to the use and utility of the theory. It could in fact be argued that explanatory capacity is a virtue, characteristic or quality which is *derived* from these other fundamental virtues.[8] In this way, van Fraassen opposes the concept of explanation as one of the key goals of science, linked in turn to its consideration as one of the pre-eminent theoretical virtues sought by scientists. In his opinion, the only minimum fundamental virtues, or the minimum fundamental acceptability criteria in the context of a theory are consistency (both internal and with the facts) and empirical adequacy; thus, we can only talk about having an explanation for something if we have previously an *acceptable* theory which explains it.

Therefore, to the extent that (a) adequacy is a *prerequisite*, since its absence opens up the possibility of inconsistencies with the facts; (b) a theory does not have to explain each and every one of the events in its domain, providing it is consistent with them; and (c) an explanation is not always required or asked for when one may be had,[9] explanation or explanatory capacity is a derived, pragmatic virtue. Consequently, it should not be understood or conceived as a relationship between theory and the world, but rather as a relationship between the theory, the world and the epistemic community (context).

From this perspective, explanation as such is not one of the goals of science, although van Fraassen admits that it may have a certain value in the achievement of said goals. Given that having a good explanation mainly consists of having a theory with the aforementioned fundamental qualities, the search for explanations mainly consists also of finding theories that are simpler, more unified and, above all, more susceptible to being empirically adequate; in other words, its value is limited to its

---

[7] Based on the fact that the basic elements which make up a linguistic situation from a pragmatic perspective are the speaker, the statement or series of statements proposed or expressed, the audience and the factual circumstances, a factor will be pragmatic if it refers to the speaker or the audience; and it will be contextual if it pertains specifically to the particular linguistic situation in question.

[8] This does not mean that possessing these fundamental virtues automatically turns a theory into a good explanation. For this, the pragmatic aspect of the explanation is also required. However, what is underscored is that "the *epistemic* merits a theory may have or must have to figure in good explanations are not *sui generis*; they are just the merits it had in being empirically adequate, of significant empirical strength, and so forth." van Fraassen (1980, 88).

[9] Examples of this situation would be the absence of an explanation for gravity in the case of Newtonian celestial mechanics, or the debate on hidden variables in quantum mechanics. In this case, the idea of empirical equivalence comes into play. If explanation were a fundamental virtue, it could be argued that when faced with two empirically equivalent theories, that with the greater explanatory capacity should be accepted. However, the scales do not seem to tilt towards placing this capacity in play, unlike that which occurs with the greater or lesser degree of empirical gain.

possibility of offering some kind of gain in the empirical results. In the absence of this gain, explanation has no great value in and of itself. To put it in a slightly different way, the search for explanation is important to science only because we pursue it through these other basic qualities. Explanation itself offers no "bonus" for these other theoretical virtues. The reasons one might use to argue in favour of said search are pragmatic reasons, and if said "bonus" is admitted, it is absolutely pragmatic, relative to the users of the theory and the context, rather than to the relationship between the theory and the facts.

8.3.1.1 The Answers to Why-Questions

With this concept of explanation as a derived and pragmatic virtue, van Fraassen develops his model based on the works of Bromberger, B. Hansson's proposal regarding the elements of contrast, Belnap's erotetic logic and his own reading of the Aristotelian theory of the four causes. In accordance with the above, an explanation is an *answer* to a *why-question*; at the same time, a question of this kind is also a request for an explanation. The general formulation of a question is as follows:

Why (is it the case that) P? Where P is a statement.[10]

However, every why-question includes, as a general underlying structure, a contrast class, a set of alternatives against which the fact about which the question is asked is contrasted; without this class, which provides the background against which the question is asked, the question would be void of content, and one would not know what was being asked. The explicit formulation of a why-question is therefore:

Why (is it the case that) P in contrast with (other members of) X? Where P is the *topic* of the question, and X *the contrast class*, i.e., the class of all the alternatives, including the topic.

The topic and the contrast class alone, however, are not enough to ensure the total identification of a why-question, because even once the contrast class has been established, different types of response may be possible depending on which relationship is deemed most relevant in a given context in order for the response to count as an explanation. In other words, the question is indeterminate until the type of answer considered explanatory in the given context has been established. van Fraassen calls this third element required to identify a why-question the *explanatory relevance relation*, R. R relates propositions (or facts), A, with topics and the contrast class, so that A will have R with topic P, and contrast class X, if and only if A is (taken in the context) explanatorily relevant for P to occur instead of the other members of the contrast class. In any case, R only determines the *type* of response

---

[10] P is a statement, but an explanation is a response, not a statement or an argument, and it is a response to a why-question, thus a theory of explanation must be a theory of said questions, of the questions which ask why. P, on the other hand, establishes the fact to be explained, the *explanandum*.

deemed relevant, not the response itself, because the context may consider various answers to be relevant.[11]

Thus, a why-question is identified through the topic P_k, the contrast class $X = \{P_1, \ldots, P_k \ldots\}$ and the explanatory relevance relation R. We can therefore represent a why-question in the following way:

$$Q = <P,X,R>$$

A proposition A is *relevant* for Q precisely when A is in an R relation with $<P_k, X>$. X and R depend on context; they are contextual factors because in the author's opinion, they are not determined either by the totality of accepted scientific theories or by that for which an explanation is requested. This continuous reference to context as a key determining factor is one of the basic characteristics of this model of explanation. Thus, for example, in the majority of cases, the contrast class, which would be included in all why-questions, is not described explicitly because all those involved in a situation of this kind are already aware of the alternatives in question. The same thing occurs with the explanatory relevance relation, and even with the determination regarding which question is actually being asked by means of the stated query: it all depends on the context.[12] In any case, although there are contexts, such as scientific contexts in periods of normal science, in which X and R may be established more firmly, in others, they are subject to a greater degree of variation. Having specified the elements which make up why-questions, it is possible to characterise their answers: The canonical way of expressing a *direct answer*[13] to a why-question Q is as follows:

(*) P_k *in contrast to* (the rest of X) *because A.*

This statement expresses a proposition, and the exact nature of the proposition expressed is determined by the same context which selected Q as the proposition expressed by the question "why P_k?"[14] This proposition makes four claims:

1. P_k is true.
2. Every other element of the contrast class is false.
3. A is true.
4. A bears R to $<P_k, X>$.[15]

---

[11] This clarification would be the result of Aristotelian teaching: the theory of the four causes establishes four types of explanatory relevance relation characteristics, depending on context.

[12] In van Fraassen (1980, 136), as examples of contextual variables, the author refers to the assumed suppositions, accepted theories, images of the world or paradigms to which subjects adhere in a given context.

[13] A direct answer is that which provides sufficient information to respond completely to the question; or in other words, a direct answer implies a complete answer.

[14] This implies that some of the same contextual factors, and specifically R, may appear in the determination of the proposition expressed by (*).

[15] When we say *because A*, we are asserting that A is explanatorily relevant for $<P_k, X>$.

In accordance with these considerations, van Fraassen (1980, 144)[16] offers the following definition:

B is a *direct answer* to question $Q = <P_k, X, R>$ if there is some proposition A such that A bears relation R to $<P_k, X>$ and B is the proposition which is true exactly if (P_k; and for all $i \neq k$, not P_i; and A) is true.

The proposition A is called the *core* of the answer B, given that the answer can be abbreviated to *"because A."* This analysis, however, requires another element to be taken into account: every question supposes certain presuppositions, and why-questions are no exception.[17] Thus, a why-question Q presupposes that:

1. Its topic is true.
2. In its contrast class, only its topic is true.[18]
3. At least one of the propositions that bears its relevance relation to its topic and contrast class is also true.

The question Q will be accepted in a given context if said presuppositions fit into the body of factual information accepted in said context. If we call this body of information K, then the question Q *arises* in the context if K implies 1 and 2 (the central presupposition of Q) and does not imply the negation of any presupposition; in other words, the acceptance of a question in a context as a question which requires an answer demands, as a necessary condition, that the information accepted in that context confirm that, of all the alternatives of X, only the topic is true, and does not, moreover, exclude the possibility of an answer existing, because otherwise the question would simply not arise — it would be rejected.[19] If Q arises in the context, then it is possible, if found, to provide an answer-explanation, which will take the form of "because A." This set of elements configures van Fraassen's explanatory model which, in accordance with our conceptual nucleus, we can sum up as follows (Díez and Moulines 1997, 249):

1. The *explanandum* is a proposition P_k (singular or general), associated with a contrast class which includes other alternative propositions P_1, P_2, etc.
2. The *explanans* is a proposition A.

---

[16] In van Fraassen (1980, 136), as examples of contextual variables, the author refers to the assumed suppositions, accepted theories, images of the world or paradigms to which subjects adhere in a given context.

[17] Indeed, if one is not willing to accept some of these presuppositions, the question is considered inappropriate, and is therefore rejected.

[18] These two first suppositions constitute the *central presuppositions* of Q.

[19] The issue of the circumstances under which a why-question arises is central for the author, since he considers the problem of the rejection of demands for explanation to be one of the main obstacles facing the majority of models of explanation. His theory, on the other hand, can account for these rejections: we reject a question of this type by saying that it does not arise in the context. For example, as W. Salmon states in *Scientific explanation and the causal structure of the world* (1984, p. 105), in Aristotelian physics, we can ask for an explanation of the uniform movement of a body, while in Newtonian physics we ask only about the change in movement. In this context, the question of explaining uniform movement simply does not arise.

3. The explanatory relation is that of *explanatory relevance* R, determined by the context: the *explanans* explains the *explanandum* if and only if A is explanatorily relevant, according to the context, so that P_k occurs instead of P_1, P_2, etc.

8.3.1.2 Evaluation of Answers

The model, however, does not only include the characterisation of explanation as answers to why-questions. So far, we have considered these questions and their answers, with particular focus on their acceptance. Within the framework of his theory, however, van Fraassen also looks at their evaluation. In other words, his theory of explanation includes two parts or theses, one relating to the determination or definition of answers to why-questions, and another relating to the evaluation of (i.e., how to assess) these answers.

Once question Q has been posed in a context set against the backdrop K of accepted theory, in addition to certain information, and given the answer *because A*, there are at least three ways of evaluate this answer. The first relates to the evaluation of the core of the answer, of proposition A, as being acceptable or possibly true. If the context of knowledge K implies the negation of A, i.e., if it says that A is false, then we reject *because A* as an answer; if this is not the case, we must then consider the probability that K bestows on A. The second way of evaluate the answer is related to the degree or extent to which A *favours* the topic of Q in comparison with the other members of Q's contrast class. And finally, the third method compares *because A* with other possible answers to the question posed. This comparison includes three different aspects: (a) whether A is more probable, in relation to K, than these other possible answers; (b) whether A fosters the topic of Q to a greater extent than the alternatives; and (c) whether these alternative answers may render A partially or totally irrelevant for the topic.[20]

If, in context K, the question arises as to why P_k rather than P_1, ..., P_i, and we must assess how and to what extent the answer favours the topic, K must imply P_k, as well as the falsity of the rest of the members of the contrast class. However, if the information regarding the truth of the topic and falsity of the other members of the contrast class were implied only by K, this is not enough (it would be irrelevant) to gauge the extent to which A favours the topic. Therefore, the evaluation only uses that part of the background information which constitutes the general theory regarding these phenomena, in addition to other "auxiliary" facts which are known but which do not imply the fact being explained. As the author himself points out: "the probability to be used in evaluating answers is not at all the probability given all my background information, but rather, the probability given some of the general theories I accept plus some selection from my data." van Fraassen (1980, 147).

Given that the most common situation in which we ask for an explanation of P_k is one in which we always know that the topic or explanandum is true and the other members of the contrast class false, and this follows on trivially from K, if we do not want to trivialise the characterisation of the explanation, we must distinguish

---

[20] Whether these responses *displace* A from the topic.

between K and a certain part, K(Q), of K[21]: only in reference to this can the answer *because A* to the question posed be evaluated.

The question, however, is that it is not easy to see or determine how this part, K(Q), should be selected: "Neither the other authors nor I can say much about it. Therefore the selection of the part K(Q) of K that is to be used in the further evaluation of A, must be a further contextual factor."[22]

Let us suppose, in any case, that in some way we have isolated this part that we can use to evaluate answer A. In this case, we can say that, in this context, A is best qualified to favour topic P_k if A and K(Q) together imply the truth of the topic and the falsity of the other members of the contrast class. In the event that this does not occur, we must assess the way in which A favours the topic in terms of the way in which it redistributes the probabilities between the members of the contrast class. If we call the probability based only on K(Q) prior probability, and the probability based on K(Q) and A posterior probability, then A will obtain this rating if the posterior probability of P_k is equal to 1. If not, then it can still attain it if it is capable of displacing the mass of the probability function towards P_k, either, for example, by increasing the probability of the topic while decreasing that of the other members of the contrast class, or by maintaining the former while reducing that of at least some of its closest rivals.

In the author's opinion, what we must consider are the minimum advantages possessed by P_k over the other members of the contrast class, and the number of alternatives in said class in relation to which the topic has these minimum advantages. The former should increase and the latter decrease. The increase in that which favours the topic in relation to its alternatives may coincide with a decrease in the probability of the topic, but the two processes are fairly compatible, meaning that the mere fact of the probability dropping is not enough to disqualify the answer; in other words, A may favour P_k even if the probability of A decreases. Moreover, there is also another way (related to Simpson's paradox[23]) in which A may provide information which favours the topic. In accordance

---

[21] To put it in a slightly different way, the aim is to reduce the framework knowledge by excluding this part of the information, but without eliminating much more. As the author says (ibid): "In deterministic, non- statistical (what Hempel called a deductive-nomological) explanation, the adduced information implies the fact explained. This implication is relative to our background assumptions, or else those assumptions are part of the adduced information. But clearly, our information that the fact to be explained is actually the case, and all its consequences, must carefully be kept out of those backgrounds assumptions," if we want to avoid said trivialisation.

[22] van Fraassen (1980, 147). This comment generates a surely undesirable, but I believe inevitable, impression for the author, namely that context seems to be a kind of bottom drawer containing a hodgepodge of all that cannot be theorised or which has proven problematic in other theories of explanation, and which is dissolved by a reference to context rather than a fundamental proposal.

[23] This paradox shows that any association – $P(A/B) = P(A)$; $P(A/B) > P(A)$; $P(A/B) < P(A)$ – existing between two variables in a given population may be inverted in the subpopulations, with a third variable being found which correlates with both. In this sense, Cartwright (1983, 25), points out that the counter-examples to the statement that causes increase the probability of their effects works in this way. Thus, this increase occurs only in those situations in which this correlation is *not* found.

with this, the author adds the following to the characterisation of *favouring* (i.e., that A favours P_k over and above $P_1, ..., P_i$): if $Z = \{Z_1, ..., Z_n\}$ is a logical partition of relevant explanatory alternatives, and A favours P_k over and above $P_1, ..., P_i$ if any member of Z is added to our background information, then A favours P_k over and above $P_1, ..., P_i$.[24]

Finally, van Fraassen considers those situations in which an answer can be rendered totally or partially irrelevant by other answers which may be offered. Here we are talking about *displacement* in the Reichenbach-Salmon sense: P *displaces* A in relation to B if the probability of B given P and A is the same as the probability of B given P alone; i.e. P renders A irrelevant.

For van Fraassen, this criterion should be applied with care for the following reasons: firstly, because it is not important that a proposition P displaces A in relation to B if P is not the core of the response to the question; secondly, because a displaced response may be partial, but not necessary irrelevant, and it may also be good despite being partial; and thirdly, because if different answers are provided to a question, some of which are displaced by another equally good answer, then the most correct conclusion, according to the author, is that if an answer is displaced by another, and not vice versa, then the latest one must be better in some way.[25]

This characterisation of the evaluation is, as van Fraassen himself points out, neither precise nor complete, but its imprecision and incompleteness do not invalidate the theory of explanation developed, according to which explanations are answers to specific, contextual why-questions. In effect, this analysis enables us to reconsider and, in the author's opinion even solve, some of the problems which every adequate theory of explanation must face.

---

[24] Adopting the example given by Cartwright (1983, 23–24), of the relationship between smoking and heart disease, van Fraassen states that in response to the question "why does Thomas have heart disease?" the answer "because he smokes" favours the topic that he has heart disease in a direct yet derived sense, because the probabilities of contracting said disease increase if you smoke, regardless of whether or not the smoker engages in physical exercise, and must be one or the other. The author also recognises that it is in the context of this second assessment criterion that both the Hempelian criterion of offering reasons to expect, as well as that of Salmon regarding statistical relevance, can be applied. Salmon (1984, 108–109), agrees that the notion of favouring should admit cases of negative relevance, but rejects the idea that only the favoured members of the contrast class may be explained, because we understand as much (or as little) about the favoured results as we do about the non-favoured ones.

[25] Using his own examples — van Fraassen (1980, 150–151) — in the first case, if we know that Paul has just killed Peter and we ask why he is dead, and the answer given is that he received a heavy blow on the head, this is no worse if it is statistically screened off by other types of information; in the second case, we know that there must be a true proposition such as "Peter received a heavy blow on the head with impact x" but this only means that an answer richer in information is possible, not that the given response must be disqualified; and in the third case, if we ask why the system is in state A_n at the moment t_n in response to a determinist process in which state Ai and no other state is followed by state A_{i+1}, then the best answers to said question may take the form of "because the system was in state A_i at moment t_i," but each of these answers is screened off the event described in the topic by another equally good answer. In this case, if the answer is screened off by another, but not conversely, then the last is better in some respect.

Earlier on in this chapter, we saw how van Fraassen resolves the rejection of certain why-questions.[26] The contrast class, for its part, enables us to resolve the problem of the explanation of highly improbable events, such as paresis, for example. If we ask why the Mayor, rather than any of the other citizens of a city, suffers from this disease, we have a true, correct answer: because he had untreated latent syphilis. However, if the same question is asked but in contrast to other members of his country club who also have syphilis, then there is no answer. In relation to irrelevances, the body of information K accepted in the context is that which provides the solution, because K excludes the fact that taking contraceptive pills is relevant for preventing pregnancy in males, or that vitamin C is relevant for curing a common cold. The contrast class and relevance relation are determined contextually; they are contextual factors without which the description of an account as an explanation of a fact or event is incomplete.[27]

In this sense, the contextually determined relevance relation also solves the problem of asymmetries, since the latter is the result of the former, which means that "(…) it must be the case that these asymmetries can at least sometimes be reversed by a change in context. In addition, it should then also be possible to account for specific asymmetries in terms of the interests of questioner and audience that determine this relevance." (van Fraassen 1980, 130.) Thus, although in the majority of contexts the height of the flagpole is relevant for the length of its shadow, not vice versa, there may be contexts in which this relevance is inverted, such as when one wants to built a tower of a certain height in order to ensure that the shadow (which the relevant element in this case) reaches a certain spot (van Fraassen 1980, 132–133, The Tower and the Shadow). In these cases, asymmetries must be reversible by means of a change in context.

8.4 Some Critical Considerations

Some of the elements of van Fraassen's theory of explanation, with its pragmatic relativisation, specifically and especially those related to the contrast class, have been almost universally accepted, and in this sense, his contribution to the analysis of explanation is considered fruitful and valuable. However, the lack of restriction of the relevance relation seems to pose a serious problem.

Kitcher and Salmon (1998, 178–190) are particularly critical in this respect. In the opinion of these authors, if no constraints are placed on R, then we have to

---

[26] See the presuppositions of why-questions.

[27] In other words, something is, or counts as, an explanation in respect to a certain relevance relation and a certain contrast class. In turn, the fact that both are contextual factors excludes the possibility of thinking that, in cases of *scientific* explanation, the determination of the relevance of possible hypotheses and said contrast class may be automatic. For van Fraassen, the fact that an explanation is scientific only means that it is based on scientific theories and experiences; the term "scientific" says nothing about its form or about the type of information put forward — this is no different from that offered when a description is requested.

accept the undesirable consequence of anything being able to explain anything else; there may be contexts in which any proposition may count as a response to any question, simply because this relation may be of any nature. It is true that van Fraassen recognises that relevance is a tricky issue, but only for the theory of logic, (van Fraassen 1980, 152–153) meaning that he does not place himself at the same level of discourse from which Salmon and Kitchen argue their objections. Nevertheless, an answer to these objections does seem possible through the notion of *scientificity*, which does in fact place some constraints on R. As we saw earlier, the various factors are only relevant if they are scientifically relevant, and of those that are, context then determines whether they are relevant in explanatory terms. In turn, when we distinguish between scientific explanations and other kinds of explanation, the only thing we say is that the former have recourse to science in order to obtain the necessary information, and that the criteria used to evaluate them are applied using scientific theories.

This response, however, is not enough for van Fraassen's critics. As Salmon and Kitcher point out, (1998, 183–185) if someone asks why J. F. Kennedy died on that particular day, and the answer views R as a relation of astral influence, then the fact that the stars and planets were in certain positions (the knowledge of which may perfectly well be derived from scientific theories) would be a scientifically relevant factor of his death on that day, in comparison with others. The question is whether the relation of astral influence is scientifically relevant. For these authors, it is evident that astrological theory should be excluded from science, and therefore this answer would not constitute an answer to (or an explanation of) the question; but van Fraassen, not wanting to have recourse to any constraint of any kind, preferring rather to use the sole criterion of scientificity, which defined as such has an enormous pragmatic charge and seems extremely socially-dependent, is unwilling to exclude certain relations as irrelevant in *all* contexts. Thus, in the opinion of the two authors mentioned above, if we want to avoid an "anything goes" theory of explanation, we need to provide an objective characterisation of relevance relations.

In my view, what again comes into play here are key differences between the commitments of each participant in the debate. Thus, that which for Salmon and Kitcher is an unacceptable consequence for a theory of explanation (even assuming that context plays a key role in determining the explanatory relevance relation), for van Fraassen is not necessarily undesirable, since explanation is not a relation between theory and the world, but rather a pragmatic virtue: "(…) scientific explanation is not (pure) science but an application of science. It is a use of science to satisfy certain of our desires; and these desires are quite specific in a specific context, but they are always desires for descriptive information. The exact content of the desire, and the evaluation of how well it is satisfied, varies from context to context."[28]

In this sense, it is worth highlighting another vital question involved in this characterisation of explanation. Early on in the discussion of this characterisation,

---

[28] van Fraassen (1980, 156). The author's refusal to characterise relevance relations objectively is absolutely consistent with his concept of explanation.

we stated that we can only talk about having an explanation if we already have an *acceptable* theory which explains the phenomenon in question. The question regarding the acceptance of a theory forces us, according to van Fraassen, to distinguish between what a theory *says* and what *we believe* when we accept it. In this sense, the epistemological commitment implied in its acceptance is merely the belief that the theory is empirically adequate. Now, in the same way we must distinguish between what a question says — presupposes — and what we believe when we ask it. In this case, and also in the same way, the commitment implied in the posing of the question is the same as that implied in the acceptance of the theory. Thus, the aim for the author is to avoid the concept of explanation from entering the field of semantics: there may be theoretical presuppositions in a question from which one could infer a commitment to their truth (in that the speaker asks a question which includes them).

This inference, however, is not legitimate: firstly because we can apply the aforementioned distinction, and secondly, because even if this were not the case, the context in which the question arises or is posed is one in which the theory is accepted, from which we can conclude that those intervening in this pragmatic situation are immersed in the theoretical image of the world and therefore speak the language of the theory. Consequently, the language used cannot be an indicator of participants' epistemological commitments.[29]

van Fraassen called his paper on explanation, and the part of his book which deals with this issue "The pragmatics of explanation." Salmon and Kitcher aim to underscore the difference between a theory of the pragmatics of explanation and a pragmatic theory of explanation, and in this sense believe that although van Fraassen offers the best theory so far of the pragmatics of explanation, the same cannot be said if it is viewed as a pragmatic theory of explanation, because, as we have just seen, it would then be faced with serious problems.

Along similar lines, Achinstein (1993, 326–344, 326–344) analyses this model with the aim of determining whether or not it can indeed be considered a pragmatic model. The analysis concludes that it cannot, while the approach proposed by the author sustains this property or characteristic, since, briefly stated, the emphasis on contextuality is not enough to make a theory of explanation a pragmatic theory. For Achinstein, in order for a model of explanation to be characterised as pragmatic, the emphasis must be placed not only on context, but also on all the other elements which made it up[30]: the people who explain it and the audience to whom these explanations are targeted. Explanations encompass two elements or dimensions that can be included: an act and a product of said act.

---

[29] Note that this argument would effectively be valid not only for realists, but also for van Fraassen himself. Where then, does the claim that this commitment is that of the belief that the theory is empirically adequate come from? For van Fraassen, realism is not an ontological thesis, but rather an epistemological one; it is not a thesis about what exists, but rather about what we are justified in believing exists, and his position in this sense, as underscored already, is that by accepting a theory we are justified in believing only in its empirical adequacy, not in its truth.

[30] As well as the body of knowledge or shared beliefs.

If we apply this definition to van Fraassen's model, the conclusion is that it is not pragmatic, either explicitly or implicitly (Achinstein 1993, 328–333). It is true that determining which question is being asked, the set of alternative hypotheses (the contrast class) and the body of shared information (K) requires recourse to context; however, firstly, and according to Achinstein, the technical terms, such as topic, contrast class and relevance relation do not seem to require the concept of an explainer or audience; and secondly, the reference and relativisation to a context is not enough to convert a theory of explanation into a pragmatic theory of explanation: Hempel's I-E explanation is also relativised to a set of background beliefs (K) which may differ in accordance with the explanatory context, but this does not mean that his model is pragmatic. And thirdly, in relation to answer evaluation within van Fraassen's proposal, the conditions established for evaluating explanations are not pragmatic either, because their applicability does not depend, or vary with, the explainer or audience.[31] Therefore, these two elements seem to be vital for any theory of explanation to be characterised as pragmatic.[32]

8.5 Some Final Considerations

We began this chapter by considering scientific explanation and, more specifically, the models of scientific explanation, as notions involving a series of commitments and conceptions (regarding theories or their relationship with the world) which determine the way in which we understand the explanation and the models themselves. Due to space constraints, I was unable to compare some of these in order to state this basic consideration more explicitly, but in the discussion of van Fraassen's model and, above all, in the presentation of some of the criticisms levelled at it, I believe I have been able to show that no model of explanation

---

[31] Once we have completed an oration-explanation, relativising it to a specific set of alternative hypotheses and to K, the conditions for its assessment do not include terms for an explainer or an audience. In this case, the reference to these concepts is not necessary to understand the meaning of the explanation or to determine whether or not it is true.

[32] To my way of thinking, Achinstein's analysis requires several clarifications, although these do not lead to a conclusion different from that reached by the author. van Fraassen specifies that the basic elements which make up a linguistic situation from a pragmatic perspective are the speaker, the series of statements made, the audience and the factual circumstances, and points out that a factor will be pragmatic if it refers to the speaker or the audience, and contextual if it refers specifically to the linguistic situation in question. It is therefore assumed that the context must include these elements which make it up, and therefore, both the speaker and the audience. The issue, then, seems to be that this inclusion is only presupposed in his model, and this is not enough, since these elements are dispensed with in its *articulation*. They are only explicitly mentioned in the case of asymmetries, specifically in the account of the tower and the shadow, where reference is made to the intentions, desires and interests of the asker and listener in order to determine the relevance relation. In any case, if emphasis is placed, as Achinstein proposes, on what is, strictly speaking, pragmatic, in accordance also with van Fraassen's definition, the objection made appears to be justified.

(not even that of the author) is metaphysically neutral. Having said this, I will now pose a series of questions which I leave open to debate.

The first pertains to the applicability of models. Salmon made no claims regarding the universal applicability of his characterisation of causality and scientific explanation to all areas and spheres of our world. Quite the opposite, in fact — he highlights the characterisation's limitation to the field of macroscopic, non-quantum, phenomena, thus accepting and assuming its reductionist nature: "I recognise that the theory I am proposing has a highly reductionist tone. It seems to me that the approach may be sustained in natural sciences, including biology, but not in quantum mechanics. I am not certain that it would be appropriate for psychology or the social sciences." (Salmon 2002, 158.)

In relation to this, van Fraassen says: "Salmon mentions explicitly the limitation of this account to macroscopic phenomena. This limitation is serious, for we have no independent reason to think that explanation in quantum mechanics is essentially different from elsewhere." (van Fraassen 1980, 122.) It is true that we may have no such reason, but it is also true that Salmon admits that his conception of causality and explanation does not aspire to universal applicability. Whether or not this poses a problem depends on the positions adopted when proposing or tackling said problems. van Fraassen's comment makes a similar point. In Hempel's case, the option was clear: the logical structure of explanation, its logical and empirical conditions of adequacy, provided a framework for explanation in the different fields of science, including social sciences. In the case of Salmon, the limitation of his model is explicit.

The question is: should we opt for a "universalist" model of explanation, or for different models in accordance with different specific fields of science? I do not believe that we can answer this question directly and definitively, but I do believe we can analyse that which underlies the aforementioned comment. For van Fraassen, the objection made is relevant because his conception of explanation is radically different from that proposed by Salmon: explanation is not about the world, it is not a relationship between scientific theories and the world, as description is, but rather a relationship between theories, facts and context. Thus, we can accept that science provides an image of the world as a network of interconnecting events. This relationship is complex, but ordered.

However, this does not mean that the terminology of cause or causality is the most appropriate for describing this image, particularly when we consider, in light of Salmon's example, that it seems impossible to offer a complete and "universal" characterisation of said image. Each scientific theory lists a range of factors as being objectively relevant in different ways, but the choice is then determined by other factors which vary in accordance with the context in which the explanation is requested. Thus, "no factor is explanatorily relevant unless it is scientifically relevant; and among the scientifically relevant factors, context determines explanatorily relevant ones." (van Fraassen 1980, 126.) If this is indeed the case, then the model of explanation derived from these assumptions may be applied to any area or scientific domain, from the macroscopic to the quantum.

To my way of thinking, this would be one of the advantages of the model. However, and this is the second question I would like to pose, I believe that, in order to be truly "pragmatic," and here I would agree with Achinstein, having recourse only to context is not enough. We also need to include the actors — the people who explain — and the audience whom they address. If we accept the Achinsteinian distinction between the *act* of explanation and the product of this act, then we see that while van Fraassen's model, just as Hempel or Salmon's one, has focused on the products of explanation and on the evaluation of these products, it has not focused on the act of explanation itself as an illocutionary act. Evidently, and to judge by that which we have maintained so far, this cannot constitute a serious objection.

However, I believe that we could defend the complementary nature of the two approaches, both in relation to that considered an act of explanation and as regards the assessment of explanations, because here two fundamental concepts come into play: that of intent and that of understanding. Explanations are human inventions which serve human purposes, and their most important use is found in acts of explanation aimed at enabling the audience to increase their level of understanding. The aim or intent of the speaker who explains is to make that which they are explaining understandable. And I would add something else: through the explanations they provide, scientists not only aspire to ensuring that their audience understands something, they also aspire to convince them of this something. Thus, explanation is a practice, the explanatory practice, which forms part of the set of scientific practices and, in particular, discursive practices which frame interactions between scientists. From this perspective, we could say that science, in this sense, and as an activity carried out by specific human beings who interact with each other, has a dialectic and rhetoric dimension.

References

Achinstein, P. (1983). *La naturaleza de la explicación*. Mexico: FCE.
Achinstein, P. (1993a). Can there be a model of explanation? In D. Ruben (Ed.), *Explanation* (pp. 136–159). Oxford: Oxford University Press.
Achinstein, P. (1993b). The pragmatic character of explanation. In D. Ruben (Ed.), *Explanation* (pp. 326–344). Oxford: Oxford University Press.
Aristotle. (1988). Analíticos Primeros y Analíticos Segundos. In Aristotle (Ed.), *Aristotle, Tratados de Lógica (Órganon)* (Vol. II). Madrid: Gredos.
Cartwright, N. (1983). *How the laws of physics lie*. Oxford: Oxford University Press.
Díez, J. A., & Moulines, C. U. (1997). *Fundamentos de filosofía de la ciencia*. Barcelona: Ariel.
Estany, A. (1993). *Introducción a la Filosofía de la Ciencia*. Barcelona: Crítica.
Feigl, H., & Maxwell, G. (Eds.). (1962). *Scientific explanation, space and time, Minnesota studies in the philosophy of science* (Vol. III). Minneapolis: University of Minnesota Press.
Gonzalez, W. J. (Ed.). (2002). *Diversidad de la explicación científica*. Barcelona: Ariel.
Hacking, I. (1983). *Representing and intervening*. Cambridge: Cambridge University Press.
Hempel, C. G. (1988). *La explicación científica. Estudios sobre la Filosofía de la Ciencia*. Barcelona: Paidós.
Hilgevoord, J. (Ed.). (1994). *Physics and our view of the world*. Cambridge: Cambridge University Press.

Kitcher, P. (1989). Explanatory unification and the causal structure of the world. In P. Kitcher & W. Salmon (Eds.), *Scientific explanation* (pp. 410–505). Minneapolis: University of Minnesota Press.

Kitcher, P., & Salmon, W. (Eds.). (1989). *Scientific explanation*. Minneapolis: University of Minnesota Press.

Kitcher, P., & Salmon, W. (1998). Van Fraassen on explanation. In W. Salmon (Ed.), *Causality and explanation* (pp. 178–190). Oxford: Oxford University Press.

Levine, G. (Ed.). (1993). *Realism and representation*. Madison: University of Wisconsin Press.

Perdomo Reyes, I., & Sánchez Navarro, J. (2003). *Hacia un nuevo empirismo. La propuesta filosófica de Bas C. van Fraassen*. Madrid: Biblioteca Nueva.

Pitt, J. (Ed.). (1988). *Theories of explanation*. Oxford: Oxford University Press.

Salmon, W. (1984). *Scientific explanation and the causal structure of the world*. Princeton: Princeton University Press.

Salmon, W. (1990). *Four decades of scientific explanation*. Minneapolis: University of Minnesota Press.

Salmon, W. (1998). *Causality and explanation*. Oxford: Oxford University Press.

Salmon, W. (2002). La estructura de la explicación causal. In W. J. Gonzalez (Ed.), *Diversidad de la explicación científica* (pp. 141–159). Barcelona: Ariel.

van Fraassen, B. (1977). The pragmatics of explanation. *American Philosophical Quarterly, 14*(2), 143–150.

van Fraassen, B. (1980). *The scientific image*. Oxford: Clarendon Press. Spanish Edition: Van Fraassen, B. (1996). *La imagen científica* (S. Martínez, Trans.). Barcelona/Mexico: Paidós/UNAM.

van Fraassen, B. (1985). Salmon on explanation. *The Journal of Philosophy, 82*, 639–651.

van Fraassen, B. (1989). *Laws and symmetry*. Oxford: Clarendon Press.

van Fraassen, B. (1994). Interpretation of science: Science as interpretation. In J. Hilgevoord (Ed.), *Physics and our view of the world* (pp. 169–187). Cambridge: Cambridge University Press.

van Fraassen, B., & Sigman, J. (1993). Interpretation in science and in the arts. In G. Levine (Ed.), *Realism and representation* (pp. 73–99). Madison: University of Wisconsin Press.

Vega, L. (1990). *La trama de la demostración*. Madrid: Alianza Ed.

Vega, L. (2003). *Si de argumentar se trata*. Madrid: Montesinos.

Chapter 9
Values, Choices, and Epistemic Stances

Bas C. van Fraassen

Abstract Naturalized epistemology, as advanced by Willard Van Orman Quine, appears to make epistemology merely descriptive in form, rather than normative. In striking contrast with the tradition, it appears to leave no place for value judgment in rational formation and change of opinion or belief. Some more recent forms of naturalism in epistemology are more liberal in this respect, but still mainly focus on instrumental value alone. The role of value judgment as it appears in epistemic and doxastic tasks faced in science, as well as in more common practical pursuits, will here be re-examined with a focus on philosophical positions characterized as stances rather than dogmas. The difference between "first-person" expression of value judgments and "third-person" attribution is crucial to the characterization of tasks involved in our epistemic and doxastic life. The conclusion advanced is that such tasks, at every level, involve value judgment, and that epistemology cannot escape involvement with the normative going beyond instrumental value.

Keywords Naturalized epistemology • Value judgement • Philosophical stance • First-person language • Objectivity

9.1 Introduction: What Are Epistemology's Concerns?

Epistemology is a discipline, directed to an area of inquiry, and I'd like to begin with an idea about just what that area is.

B.C. van Fraassen (✉)
Department of Philosophy, San Francisco State University,
w/n, 94132 San Francisco, CA, USA
e-mail: fraassen@sfsu.edu

W.J. Gonzalez (ed.), *Bas van Fraassen's Approach to Representation and Models in Science*, Synthese Library 368, DOI 10.1007/978-94-007-7838-2_9,
© Springer Science+Business Media Dordrecht 2014

9.1.1   Three Levels of Doxastic Tasks

There are a variety of mundane tasks both in science and in daily life that aim to form, fix, or change opinion and belief. I use the term "opinion" generically and very broadly to cover any kind of assessment or appreciation of a situation or condition.[1] The example traditionally chosen as paradigm in epistemology was judicial inquiry, though in recent centuries as often as not that is overshadowed by that of the scientific one. The judge and jury in a trial "weigh" the evidence and aim to arrive at conclusions beyond reasonable doubt. In our little ways we do this sort of thing every day; and in their grander ways, so do scientists. What we arrive at is in general a complex of assessments, judgments, attitudes toward, and even feelings about, that situation or condition. Rarely is the outcome anything like a single-sentence conclusion.

But at a level just above those mundane assessments, so to speak, there are the tasks of evaluation of that very process itself. This too we can see before looking at philosophy:

> Thus, in Pennsylvania at least, the trial judge regularly gets to file a post-appeal opinion that often reads like the trial judge's own brief in support of affirmance. But if the trial judge … concludes [for example] that the order being appealed … represented an improper exercise of discretion, then the appellate court can entirely ignore the opinion. (Bashman 2007).

While the questions raised in epistemology do pertain quite often to what I called the mundane tasks, they pertain more often to that second level of assessment.

Note though that reasoning on that second level is still all part of the quotidian "enterprise of knowledge," the enterprise that aims to place us in a position in which we can feel at home in the world and know our way around in it. The appellate judges do not need additional teaching from philosophers; neither do the peer reviewers for scientific technical journals.

So peculiarly philosophical questions enter only when we attempt to reach a synoptic vision and understanding of this entire enterprise. That is not to exclude that epistemology can bring useful insights also to judges and scientists. But the insights the philosopher seeks are not the same sort as those the aspirant judges or experimentalists hope to gain from their mentors. If we are to locate epistemology in this landscape, as a reflection on this enterprise, we must accordingly map out still a third level. Philosophers have their own vocabulary for framing those first and second level assessments I described:

> Were the conclusions consistent and in coherence with the evidence? Proportionate to the evidence? Arrived at rationally and — as virtue epistemology emphasizes —, conscientiously,

---

[1] As the quotation in the next paragraph shows, "opinion" is also a technical term in law. "Appreciation," a term I should like to introduce into epistemology to widen "opinion," is similarly a technical term already, both in art criticism and in military intelligence (as used in Sheffy 1990). The first two entries in the Oxford English Dictionary for "appreciation" favor the meaning here indicated, though with a nod to the currently dominating connotation of approval, which I do not mean to include (and is not included in the military use).

responsibly, with due attention to the epistemically accessible alternatives? Was the selection of something as evidence itself ratifiable?

And so forth.

9.1.2   The Ostensible Role of Value Judgments, and a Rival Conception

These questions ask for value judgments. Some of the values involved are specifically epistemic, but others are clearly of much broader span, and adapted from ethical judgment. If it is concluded that, as in the above example, a court's decision was reached through an improper exercise of discretion, then the assessment of this "mundane" epistemic task clearly mobilized value judgments going well beyond questions of accuracy, and seem indeed to shade off into ethical concerns.

Conceived this way, epistemology appears to have value judgments at the center of all its concerns. For so described, the enterprise of knowledge involves a constant practice of value-judging, in the course of forming, fixing, and changing opinion.

But that is not the only conception in play today, and I want to begin with a defense against a rival conception that targets the apparent role of value itself in our understanding of that enterprise of knowledge. *What if the rational evaluation required does not really involve value judgment, properly speaking, at all? What if it is just a matter of factual judgment, what if that is the only genuinely applicable category?* This is the fascinating if disturbing possibility offered by certain philosophical positions seen at one end of the spectrum of naturalism in philosophy.

Today most everyone in analytic philosophy seems to want to be called naturalist, and I won't shear them all over the same comb. Perhaps I am speaking only of an extreme, but it seems to me to be an extreme that imparts flavor to the whole. I will take Quine's famous "Epistemology naturalized" (1977) as my main example, and Ronald Giere's (1985, 1988) naturalized philosophy of science as qualifier.

After critique, however, a more constructive note will be in order: how we conceive of epistemology must naturally affect our entire understanding of philosophy, and we must try to see how it does.

9.2   Epistemology Naturalized – Crypto-Positivist?

In "Epistemology naturalized" Quine (1977) presents a quick sketch of orthodox, foundationalist empiricism as developed and also destroyed *malgré eux* by Hume and Carnap. Then he calls *that* "epistemology," and pronounces it a failure. All that is left behind, when epistemology is thus shorn of its false hopes, are questions for cognitive science. From this a natural conclusion follows: the only genuine questions that can be extracted from disputations in epistemology are questions of fact, which fall within the scope of empirical science itself.

9.2.1 Diagnosing Quine

What exactly does Quine advocate here? Quine is conventionally typified as a scourge of logical positivism. But it is hard not to see a parallel in his (1977) to something notoriously positivist. Epistemology had traditionally concerned itself with the evaluation of belief and opinion: *is it knowledge, is it rational, is it reliable?* Each of the main terms in these questions has, at first blush, an endorsing sense: these are ways in which opinion can be *good*. Quine concludes in effect that the only significant core of such putative value judgments is a certain factual component, within the scope of (cognitive) science itself.

Perhaps this contention is part of a certain thesis about value in general. If so, the thesis was not new; it was already announced by the logical positivists a long time ago. A. J. Ayer brought the message to England in the mid 1930s:

> in so far as statements of value are significant, they are ordinary "scientific" statements, and in so far as they are not scientific, they are not in the literal sense significant. (Ayer 1946, 102–103)

We can imagine that dissection of statements applied here. I evaluate certain opinion as reliable; being an evaluation, that is a value judgment. Then the retort, presumably by Quine as well as by Ayer, is that reliability is simply a matter of fact, for "reliable" takes its meaning from e.g., correlation between the truth-values of statements issuing from this opinion and the phenomena described in those statements.

Let's introduce a name for the case in which the logic of the word classifies it as evaluative, but the basis of evaluation is simply a scale provided by a physical quantity: *factual-evaluative*.

Thus "tall" and "hot," which grade objects on a scale of height and temperature respectively, are factual-evaluative; and on the above account, so is "reliable."

The term "reliable" so understood does not imply "good." But whatever *extra* there is in the judgment that the opinion is good, the part not implied by the judgment of reliability, is the part classified by Ayer, and in effect by Quine as well, as "not in the literal sense significant."

The contention common to Ayer and Quine is therefore that upon proper disinfection, sterilization, and regimentation, all questions concerning value are systematically eliminated in favor of factual questions, from which any but factually-evaluative terms are absent.

9.2.2 Can Epistemic Norms Be Eliminated by Means-Ends Assessment?

In what sense, and to what extent can evaluative judgments be replaced by factual judgments? An example within the domain of epistemology would be the evaluation of a methodology that consists in a policy for revising opinion (including belief in theories or hypotheses) under a range of contemplated, possible circumstances. The evaluation might be *comparative*, for example, a judgment about whether Bayesian methodology is better or worse for sociology or psychic research than orthodox statistical testing. Or it might be *qualitative*, yielding only the judgment that a given methodology is minimally satisfactory, or at least rational, to adopt in a certain branch of science.

The words "better," "satisfactory," "rational" are all at first sight evaluative. But notice that in each case there is a "for," a specific purpose to which the evaluation is mentioned as pertinent. This applies even to "rational:" there are concepts of rationality which have no truck at all with absolute standards. Giere (1985, 1988) uses the term *instrumental rationality* to refer to evaluations of effectiveness of means with respect to given goals, under given circumstances that include a specification of all relevant resources.[2] A judgment of instrumental rationality is purely evaluative, for it says, for example, that methodology X is rational. But whether the judgment is correct, depends on a factual question: namely, whether X is an effective means to certain ends under certain circumstances. It appears then that by adopting the notion of instrumental rationality as the evaluative standard, we have effectively eliminated evaluation as such in favor of factual judgment. "Rational," so understood, is also factual-evaluative.

I chose the example of instrumental rationality, in order to put us in the most difficult place, when it comes to values, to resist naturalism. And I mean to show that even there, value judgment, as distinct from factual judgment, is inescapable. Certainly it appears that in the case of instrumental value judgments each request for an evaluation is effectively replaced by one for a factual judgment, and that then, if we ignore what was replaced, there is no loss. If that is so then the replacement is indeed the only valuable — perhaps the only meaningful — part of the original. But this seems to me deceptive.

9.2.3 *That the Norms Are not Eliminated*

Two points should give us pause, when considering the above "elimination" of evaluation by "factual" grading.

The first is this. Suppose by way of contrast that we adopted one or other rival conception of rationality, for example the criterion that rationality consists in being in accord with certain criteria claimed to be *a priori* norms of practical reason or the like. Then the judgment that methodology X has this feature of *a priori* rationality is also *correct if and only if* a certain factual statement is correct (namely, that X satisfies those criteria)!

---

[2] Foley (1993, Chapter 1) advocates an essentially similar account, though in much greater detail, for rationality in general. His discussion provides a good general framework, it seems to me, for Giere's proposal.

It follows then, surely, that this apparent elimination of the normative is not due to the peculiarities of the instrumental concept of rationality. It is due simply to this logical point:

> *whether or not an evaluative judgment is correct* is itself a purely factual question, once a precise standard of correctness has been specified.

This applies to *all* evaluative judgments, including the ethical! But from this logical point it does not follow at all that the value judgment can be identified with the factual statement which says that it is correct. Not only does this point not give any special support to a naturalist position over against other views of rationality, it does nothing to support the idea that judgments of instrumental rationality are — even "at heart" — anything but blatant value judgments.

To bring out the second point we can begin with quite ordinary examples of factual agreement unaccompanied by agreement in value judgments. For example, two people might agree that a certain kind of legal treatment of prisoners is cruel or humiliating — by explicitly stated factual criteria on which they have agreed — yet one may advocate, endorse, or positively enjoy it while the other is resolutely against it. Perhaps the one says it is deserved, while the other says that it is beneath us to administer such treatment to a fellow creature, regardless of desert. Or perhaps the former just says "yes, it is cruel; but so what?."

In such a case, there is complete agreement about what they take to be the relevant straightforwardly factual details, yet one stands for, the other stands against, the practice. Can we see that difference as somehow a deeper factual disagreement?

Undoubtedly a new set of factual agreements and value disagreements may enter at this point: there is no dispute about what the captive has done, nor about our ethical self-conception as noble beings, but ... And here, treacherously and confusingly, the overriding value judgment is very often framed in a putatively factual form: *Non c'è pace senza giustizia!* or instead, *Non c'è giustizia senza compassione!* Yet it is clear that these statements are made, not to show that the other is ignorant or forgetful of certain facts, but to express the differing attitude, just more forcefully.

Recognizing this, we are on the way to the conclusion that I wish to submit:

> The consideration of any evaluative category, including instrumental rationality, *loses its point entirely unless* there is also activity which does not consist purely in factual judgment.

The ends with respect to which methodology X may or may not be effective *must be adopted, must be someone's ends*, or the question of its instrumental rationality is moot.

True, whether certain ends are someone's ends is also a matter of fact. But adding this fact does not turn a factual judgment into something that can play the role of a value judgment. We may, in purely impersonal fashion, describe a range of possible ends, and ask about the effectiveness of X with respect to each. But even then, for our activity to be what it purports to be, we must have adopted some ends ourselves. Otherwise we can just react with "So what?."

We can even generalize this point to the utterances themselves, contrasting them with the activity in whose service they are made. As beautiful sounds and writing those utterances may fall short, as giving us private pleasure they may be a great success; but their *relevant* success will be defined by our goals. Unless certain cognitive ends are *our* ends, and thereby bring certain norms into relevance, our statements are mere phonetic display.[3]

9.2.4 A Diagnosis of the Tension Between Value Judgment and Its Factual Correlate

I just claimed that if we add something like "To achieve Y is K's end" to "X is an effective means to achieve Y" we still do not have a value judgment. This seems clear to me, since you and I could agree on these two facts and still disagree on whether it is a good thing to adopt means X. However there is one sort of instance where it is not only unclear but even looks wrong. That is when "Ks" is "my," when I say, in first person indicative form, "To achieve Y is my end." Setting aside for a moment possible conflicts among my ends, and rivals for maximum effectiveness for X, it seems I can't then disagree that it would be a good thing to adopt X.

This is the beginning of the diagnosis of the special character of value judgments, it seems to me. "I am in Ferrol" is a self-locating statement, a self-ascription of location. *Value judgments too are self-locating self-ascriptions.* They differ from the "I am in Ferrol" type in that they locate the judge not (only) with respect to a map or model of the factual landscape but with respect to a tabulation of value scales or standards: "(this is where I am, and) that is my/our standard among those which can apply to this place."[4]

The same correlation between facts and values that we noted above can be seen here, but here it is not pointing to a reduction. In the case of such a sentence as "I will … I aim to … I have a goal …I value …," the very same words can play two roles. The first role is that of an autobiographical statement of fact. That is different from their role when they are used to *avow*, *affirm*, or *express an evaluative propositional attitude*, or the commitment to those values. "It is the overriding goal of all my strivings to achieve self-mastery:" you can imagine this being said to express precisely what the speaker values. But you can also imagine it said by someone speaking to his therapist, and adding "and I have come to think that this is pathological," to which the therapist might then agree. In the latter context the statement is factual autobiography, but it is not — or no longer, or at least not at that reflective moment — an avowal of the speaker's values.

---

[3] This is part of a crucial more general point in Putnam 1982.

[4] Despite the phrasing, I think this point does not rest on a sharp fact/value distinction. There is no reason to deny that many of the predicates we use resist being classified wholly on either side of that divide.

To sum up then, there are two sides to the contention that values and norms can give way to factual means-ends assessment. In one way, normative or evaluative questions are *always* automatically eliminated in favor of factual questions as soon as they are made clear: i.e. when factual criteria are provided. But in another way, normative or evaluative questions *never* disappear, are never eliminated. The project to naturalize reason leans on a banal truth to insinuate a false conclusion.

9.3   How Epistemic Value Judgments Function in Dialogue

So we have seen that even the insisting that epistemic values are instrumental would not by itself remove us from the realm of value. But if value is not lost altogether, in that case, is it made debilitatingly relative? If the extra beyond the factual core is merely, for example, an expression of personal or communal preference, the ingression of values into the domain of epistemology would seem as destructive of epistemology as anything the naturalizing project could do. We would also not be much farther from the crypto-positivism to be suspected in that Quinean turn. Reducing value judgments involved in assessing belief and opinion to personal or even communal preferences is not much to be preferred over eliminating them altogether.

On the other hand, there is not much gain in postulating "objective values" existing somewhere in their own splendor, looking down up on our human muddling. Hearing of them, the human muddler can once again just say "So what?" For if the values are not our own, they have no motivational role for us.

If these two possibilities were the only ones — either mere personal preference or "objective values" disjoint from our own striving — we would indeed be caught in a sort of reductio ad absurdum. But such dichotomies are generally as unreal as they sound, and we should hope to see our way through them.

9.3.1   Exploring the Positivist Precedent

We can explore this by looking a bit further into the logical positivist precedents for naturalized epistemology, and how it relates to positions on ethical values.

When A. J. Ayer brought the logical positivist gospel to England, he made one very prudent rhetorical move. He wanted to give pride of place to scientific ideals of rationality, among all values. But he also wanted to promulgate the emotive theory of value. So he presented the special case of rationality *first*, and the general one of value afterward. In that way he could sound commendably *naturalistic* about scientific rationality and dismissively *relativist* about value in general. Scientific rationality might be instrumental rationality, but was very important, and to insist on that was certainly not supposed to be a matter of merely emoting!

Yet he could not stray too far from the account of value in general, and so when it comes to rationality Ayer opts for preferences that are communal. His account of rationality was that of instrumental rationality as gauged by the person's or community's own lights:

> to be rational is simply to employ a self-consistent accredited procedure in the formation of all one's beliefs… For we define a rational belief as one which is arrived at by the methods which we now consider reliable… And here we may repeat that the rationality of a belief is defined, not by reference to any absolute standard, but by reference to our own actual practice" (1946, 100–101)

9.3.2 Enter the First Person, Once Again

This is one place where we feel acutely the strain between first-person and third-person points of view. We have little difficulty agreeing to Ayer's statement (ignoring the implications of "simply," "define," and "defined") when the "we" and "our" refer to a community which we at the same time *belong to* and *endorse* — for example the scientific community whose values and ends we *underwrite* as our own. Ronald Giere's characterization of instrumental rationality as "using a known effective means to a desired goal," "policy… based on solid empirical findings about effective strategies for pursuing various scientific goals" (1988, 7 and 10) coincides with Ayer's at least in that special case.

But here is the problem: what do these views entail for judgments of rationality that cross community boundaries? Do we evaluate others' beliefs and practices as rational by reference to their actual practice, or by reference to ours? If the former, we are in a bind if we do not endorse what they endorse. If the latter, we may judge as irrational exactly what they, if following Ayer in this, will have to count as the only course that rationality allows. Aren't we facing here a dilemma between value relativism and value imperialism?

The view of rationality that emerges from my two quotations of Ayer is the following. The judgment that a person, policy, action, or belief is rational is an evaluation, and therefore has two parts (aspects, sides). First of all it expresses the judger's attitude, his "ranking" of the object in question. To this extent the judgment is not a statement of fact, indeed by Ayer's lights, not a statement at all, and "not in the literal sense significant." But secondly, it also expresses a factual assertion, to the effect that a certain procedure is e.g., self-consistent, accredited, reliable. The connection between the two is this: (1) such terms as "accredited" or "reliable" are elliptical, and tacitly refer to certain standards; (2) when these standards are *our* standards, our evaluation ("ranking," evaluating attitude) is based exactly on the extent to which the object meets those standards.

But when are the relevant standards our own? The simple case is the one in which actor and judge belong to the same relevant community, sharing the same relevant standards. In that case, if Peter says that Paul is rational, and explains that Paul uses a self-consistent, accredited, reliable procedure, there is still an ambiguity: consistent, accredited, reliable by whose standards? Peter's or Paul's? In this case, however, when Peter and Paul belong to the same community with shared values, the ambiguity does not matter, for the standards are shared. What if they are in different communities, and the standards are not shared?

9.3.3   Three Options for Third-Person Attribution

We can discern three possible views on how the third-person attribution is to work in that case.[5] The first, which is arguably Ayer's own, is that the judge ignores the actor's standards, and makes tacit reference only to his or her own. If I say that the Romans built bridges *rationally* then I mean, according to Ayer, that they used "methods which we now consider reliable."

For an ethical example we could take "The Roman practice of slavery was immoral." This statement, which we may well make, could quite appropriately continue with "regardless of what they themselves believed about it and regardless of its function in their society." In this case the reference is surely to our own actual standards, not to the Romans.'[6]

The problem comes with a negative judgment. We would like to distinguish between the Romans building their bridges irrationally and being factually mistaken about means and resources. The judgment that they did not use methods which we now consider reliable would, on this view, imply that they did not build bridges rationally. But that surely does not follow: they used the very best methods that were available to them, what could be more rational?

It is possible that the fault lies not with the focus on the judge's own standards, but on how Ayer explicates the standards of rationality. Perhaps if we say that the Romans built bridges rationally, we are using a value term which "in itself" (so to speak) refers to standards of the subject of attribution. That would be so if "rational" means something like "not hampering oneself by using methods that one oneself regards as less than optimally available."

But this suggests the second view, that the relevant standards are those of the judged community; though it is hard to see how that could be tenable at all. Let us attribute it to the fictional philosopher Ayer*; it is the view that the judge makes reference only to the actor's standards. Hence if I say that the Romans waged war *rationally* then I mean, *secundum* Ayer*, that they did so by procedures accredited and regarded as self-consistent and reliable in their community. This is a concept of rationality as pure coherence. Indeed, it deletes from consideration all factors except the actor's opinion. For what is accredited or regarded as reliable will depend on their own opinions about what their goals are, rather than on what those goals really are (if there is a difference). Similarly it will depend on their own opinion of what

---

[5] This analysis will be reminiscent of the disputes between contextualists and invariantists in contextual epistemology.

[6] There is an associated somewhat more sophisticated view, that I call *Cosa Nostra* Ethics. The view that *what is good = what meets our own standards* faces a problem with the statement "If we had lived in Roman society then slavery would have been good" For certainly, if we had lived in Roman society then that practice would have been in accord with our standards. To overcome this objection, the more sophisticated version insists that the words "our own" are to be understood as "our own actual," where the indexicals are taken to reach outside the sentence as a whole to the speaker's context. Compare how the (coherent) *wish that I had more money than I actually have* compares with the (incoherent) wish to have more money than I have.

means are available, rather than on the means truly available. The advantage of Ayer*'s view, at least to Ayer, might be that the evaluative aspect of a judgment of rationality is utterly diminished in importance: only the facts about the actor matter at all. Or so it seems.

The reason why this second view just doesn't seem tenable in general at all is that when I gauge X's rationality, I *must* ignore some of X's own judgments. For example, I must ignore his opinion about whether or not he is rational, and even to some extent about what means are effective. I do look at his opinions about what his goals, means, and resources are like, and see if *those* entail — in my eyes — that the means are effective. As an example of what otherwise would go wrong, consider the dialogue

> Peter.   I don't fasten my seat belt, because I believe that my safety does not require that.
> Paul.   What if you had a collision with your safety belt not done up?
> Peter.   That is unlikely, given that my safety does not require fastening my seat belt.

If Peter cannot marshal opinions about seat belts and accident statistics independent of his first expressed belief, we will not judge him to be rational.

There is a deep-going difficulty for Ayer*'s construal. Not having a foundationalist epistemology, we cannot think of Peter's estimates and probabilities as determined uniquely by a basis of evidence uninfected by his own epistemic attitudes. If we look for "basic" beliefs in Peter's opinion, some of them may be assertions of safety, reliability, and effective means. His probabilities should be coherent with each other, and cannot be "taken apart" even to *that* extent.

In this example, I would still judge Peter to be irrational even if I thought his opinion was coherent. My explanation would consist in *my selecting* certain of his opinions as "his evidence," and evaluating how his opinions go beyond *those* by my own standards for prudent extrapolation and risk assessment.

It may be remarked here that I am pointing to a critique of Peter's factual opinions rather than of his standards. But that point is moot since, as I remarked, Ayer*'s construal leaves no room for anything else. If all that matters in the assessment of Peter's rationality is how his conduct looks by his own lights, then the only remaining target is his *opinion* about his conduct.

The third view, which we can now formulate, may give us the proper balance. It is more easily read into Giere's words than into Ayer's, and perhaps it was actually Giere's view. Let us refer to a person's goals and opinions together as his *point of view*. Then this view is this:

> a judgment of rationality is an assessment relative to the judge's point of view, based on a selection of his own values and opinions, of a relationship between the actor (or actor's conduct, policy, opinion, beliefs) and the actor's point of view.[7]

As an example, this view would apply to our judgment about slavery in Roman society as follows. We judge factually that in the Roman's point of view, slavery is an efficient means to certain social ends, and that the effect on the slaves is an

---

[7] The point about selection must be included; if omitted we fall back into the first view which will effectively reject all other points of view.

irrelevant factor. Whether or not we agree with the efficiency of those means, we judge the Roman practice as immoral because it violates our standards about which effects are to be taken into account as relevant.

After the difficulties with Ayer's and Ayer*'s one-sided views, this view is to be recommended. Spelling out the relationship in question is not easy, however. The evaluative judgment will, on any given occasion, rest on a selection from the judge's goals and opinions on the one side, and on the other side a similar selection (by the judge) from the actor's point of view. The attitude expressed derives from the judge's point of view *plus* his selection (surely one with a good deal of leeway) of opinions and goals of the actor (to which to relate the actor's *other* opinions or goals or the actor's conduct, etc.) There is a very strong element of subjectivity here, in the sense that the judge's point of view intrudes crucially in this value judgment. But that is what makes it a value judgment, an evaluation.

If this is correct then reason can indeed not be naturalized, at least not to the extent that questions of value can be effectively removed from epistemology. On this view, assessment of instrumental rationality too is a value judgment. What the judge's own point of view is, that includes what his goals — his ends — are, and so his judgment evaluating the actors' rationality will be in part an expression of what he, the judge, endorses. The assessment of rationality is then also the expression of an attitude which exists at all only because certain standards are — as the judge would say — *our* standards, the ones we personally or communally adopt or endorse, and which define *our* commitments.

If we accept this third view of judgments of instrumental rationality as correct, then we cannot stay with naturalized epistemology. So interpreted, Giere's proposal takes us abruptly out of the Quinean view. What Quine advocated as naturalized epistemology, such as research into actual scientific or common practice, is not an activity to be discounted; but it doesn't need philosophy to encourage it. Epistemology, on the other hand, focusing in part on what it means for policies (for scientific research, for change of opinion) to be rational, involves itself in an issue which cannot be equated to any value-neutral, point of view-less, commitment-free question.

9.4   Stances as Involving Values and Inconceivable Without Them

So, with the naturalist rival out of the way, let us return to the beginning, to the subject matter of our discipline. I described epistemology at the outset as addressing and reflecting on a variety of mundane tasks both in science and in daily life that aim to form, fix, or change opinion, using the term "opinion" to cover any kind of assessment or appreciation of a situation or condition. I added that even there we can already discern two levels, the second engaged in critique and evaluation of the first. As I also emphasized, the appreciation arrived at, on either level, is almost never something very simple: the "conclusion" is a cluster of judgments, both factual and evaluative, and attitudes of various sorts toward the topic of concern.

In the case of a jury, it may seem that what is arrived at is a simple factual judgment and nothing else: "guilty" or "not guilty." But what the jury's cognitive process ended with, and indeed aimed at, was a determination to convict or not convict, and this through a process of evaluation during which the epistemic attitudes toward evidence and accused evolved rationally. In the case of a literary critic or art critic, we see more easily the evaluation arrived at, even if the critique is not a review, and even if it does not pretend to arrive at a final judgment as to merit. In a less professional context, the appreciation we reach of a situation — such as a new place to live, a climbing route, a colleague, a family, a club's morale — the cluster of attitudes is even more obviously complex: it springs to the eye here that the "outcome" involves commitments, expectations, hopes, affections, orientation toward the future.[8] Hence the outcome of such a deliberation, issuing in a new appreciation of the situation, is indeed a cluster of attitudes and judgments, of various sorts, factual appreciation and evaluation. I have adapted the term "stance" to refer to such a cluster (van Fraassen 2002).

9.4.1 That Philosophical Positions Instantiate the Very Same Form

All of this applies to the management of opinion, broadly conceived, which is the proximate subject of philosophizing in epistemology. But the same conclusion applies, mutatis mutandis, at the level of epistemological reflection as well. There we see various philosophical positions, distinct epistemologies; what are such positions like? As I have argued elsewhere, as a cluster of attitudes, commitments, and values as well as beliefs, is also how we must in general describe a philosophical position. Hence a philosophical position too is a stance, in that in general it does not consist in a factual thesis alone, but in such a cluster of attitudes.

To take a philosophical position involves typically some factual opinion, that one must be prepared to defend. Here especially it is important to understand "opinion" in its broad sense, that does not exclude any other mode of appreciation; and those other modes are essential to what the position is. Taking a philosophical position involves taking on attitudes that include such propositional attitudes as belief and doubt, but certainly other sorts of attitudes as well, which we can only describe by such words as "accept," "demand," "reject," "disdain," "defer," and yes, "value."

If we felt a threat from the side of value skepticism or relativism in the discussion so far, that pales beside the threat of a debilitating relativism in epistemology itself, which can now come to the fore. So let us see what challenges may have to be confronted here.

---

[8] Compare the striking descriptions of encounter, whether with a work of art or with another person in Chapter Two of Nehamas 2007.

9.4.2   Chakravartty's Three Levels

Anjan Chakravartty (2012) outlined a view of what an epistemic situation must be if
we accept the above way of understanding philosophical positions in epistemology.

There are three levels he says. These levels are indeed loosely correlated to the
levels I described at the outset. We can think of Chakravartty's levels as pertaining
to the epistemic state of an imagined person — call her Alice — with the limitations
on that person's insight, intelligence, and memory set aside for now. The first level
has the family of factual propositions that are possible objects of belief or opinion.
I take it that we can think of this family as delimited by a language, Alice's "basic"
language for factual description.

The second level is that of epistemic policies.[9] Chakravartty gives as example the
idea that one should deem explanatory virtue an important desideratum in determining what to believe, or that the methods of the sciences should be privileged over
others. These are policies regarding the generation of factual beliefs, and not themselves objects of belief: they are adopted or rejected — either is a commitment to
proceeding in a certain way. As he points out, the commitment to a policy has
accompanying opinions that come logically in train. Specifically you can't rationally adopt a policy while asserting that it is much less likely to be vindicated than
an available rival.[10] So when it comes to having opinions and policies, there are
criteria of coherence for their combination as well as for them separately.

But finally, there is a third level, which is perhaps noticed only by philosophers.
A choice between epistemic policies will not be random, nor can it be compelled on
purely logical grounds. So how to choose?

Let us call that *Chakravartty's question* for our character, Alice. It introduces
implicitly the very general question: what sort of context is presupposed for rational
choice among epistemic policies?

9.4.3   Challenges on the Third Level

Couldn't one just say: choose the epistemic policy which, according to your prior
opinion, has the best chance of being vindicated? That would be good advice, except
for two things.

First, a point where one faces a choice with respect to epistemic policy is likely
to be one where one's prior opinion, arrived at by prior policies, is in trouble. That
seems to be just the sort of moment when such barefaced reliance on one's prior
opinion fails to look like good advice. It seems rather to be the sort of point where

---

[9] The word "stance" is in danger of sliding into debilitating vagueness, so I won't call these stances,
even though I recognize that one can quite aptly say something like "her stance on questions of
evidence and theory choice consists in the following epistemic policy," cf. Teller 2004.

[10] I take it that criteria of vindication take cost into account; an almost impossible policy to follow
might be rejected despite its superiority on other counts.

we are weighing the possibility of repudiating some prior opinion, and perhaps especially opinion about what leads in general to epistemic vindication.

Second, we can readily admit that at any point, our prior opinion, plus new information to be accommodated, is all we ever have to go on. But that cliché does not settle anything! Prior opinion will include "second order" opinions, including evaluations of the reliability of ways of managing opinion. At a point where it would make sense at all to speak of choice between epistemic policies, there must be reasons to look unfavorably on the way one has been managing one's opinion in comparison with other ways that one has spurned theretofore.

So it appears that on this third level, if there really are such choices and decisions that affect one's epistemic life, we could obtain guidance only from conceptions of what epistemic policies should be like. That means: from stances that are not *epistemic stances* narrowly conceived but *stances in epistemology*, the cores of certain rival philosophical positions.

These stances will, as stances do, consist in a combination of attitudes, some of them factual beliefs and some of them value judgments, some perhaps also intentions, goals, commitments, … But we certainly see at this level a rivalry of alternative stances. So this is where Chakravartty then raises the question about stance relativism. What possible basis could there be to dispute stances at that level? If philosophers disagree in their fundamental value judgments, will they not be precluded from having any way at all to enter upon reasoning to settle their disagreement?

Let us emphasize that the impasse, if there is one, results only at the far end of an imagined dialogue among philosophers that has remained unresolved after all arguments have been settled, while some discordant value judgments still remain.

What attitude should the occupant of one stance take with respect to rival stances, on this level? Our previous ideal person Alice may be assumed to have one of these stances so let us continue with her as example. How is she to choose?

That is an unanswerable question if Alice exists before the start of human history, or if she is in the despairing position of someone lacking trust in herself to the extent of Descartes' fiction of the total skeptic. But only if. That would be a point where she has no basis of her own to proceed in any way at all. That is so unrealistic a fiction, of the ultimate information-less and value-less prior appreciation, that we can set it aside. That the question is unanswerable in that case is a logical point, but not a realistic concern.

So consider Alice, who has a stance of her own. As I see it, step one for Alice is to acknowledge the initial or logical viability of rival stances, and step two is to criticize them on the basis of elements of her own stance. Note well, in view of our earlier discussion (in which a view of Ronald Giere was adapted), that this critique will not automatically imply that Alice classes them as irrational. For she, as judge, will follow this pattern:

> a judgment of rationality is an assessment relative to the judge's point of view, based on a selection of his own values and opinions, of a relationship between the actor (or actor's conduct, policy, opinion, beliefs) and the actor's point of view.

So Alice will assess, in terms of her own philosophical stance, the relationship between the other philosopher's philosophical conduct and his philosophical stance. Her own values will include more than mere consistency, so that the other's approaching philosophical problems consistently with his own stance will not guarantee a positive judgment. At the same time she is being properly selective; and anyway, her own values may include tolerance of other points of view (in some way, to some extent); thus neither is a negative judgment guaranteed.

Alice's critique, if carried out well, will be telling for those rivals who share those particular elements of her stance, of whatever sort they may be. Other rivals who do not share them must wait to be addressed differently. Logically speaking there is no end to this, but neither is there any point in speaking in a historical vacuum. Within our historical dialogue not all logically possible rivals are there to be confronted, nor do such historically conditioned beings as we are have access to possibilities beyond our historical horizon.

Returning to question "But how is she to choose?," we can now see that this question is moot, and certainly distinct from the question of how Alice can non-trivially judge other stances. The most important point here is that acknowledging the values of others *does not imply, nor bring in train, relinquishing one's own*. Alice is proud of her values, and has the right to be, and criticizes others from her point of view, while she in turn submits to relevant critique only on the same terms. At the same time, it is not ruled out that her positive assessment of some other point of view may be followed by changes in her own.

Still, the air of complacent common sense, in this account of Alice's position, may rankle! In allowing the emphasis on shared values as precondition for reasoning about values, haven't we just succumbed to the charge of a debilitating relativism?

I do not think so, for as I see it, this subject of relativism is full of seductive false contrasts. The last task that remains here is to note these. Bringing them to light will be, by itself, already almost enough to dissolve their threat.

9.4.3.1 The False Contrast Between Factual and Value Disagreement

[Contrast 1] If we disagree about the facts then recourse is in principle simple and clearly to be had: the facts themselves will settle who is right. But there is nothing similar "out there" to settle who is right when things are differently valued.

This putative contrast is phrased in terms of ontology, apparently presupposing some view on what facts are, what there is "out there." Bracketing disagreement about such matters, let's better look into what sort of argument could support this contrast.

Why might we have the impression that disagreement on all but facts and factual judgments is impervious to reason? It is typically pointed out here that if we disagree in our value judgments, for example, then we can each make our point only by bringing up other more basic value judgments, hoping that those will be shared.

Since there is no guarantee that we will be able to fall back on ultimately shared value judgments to provide a common framework for reasoning about our differences, there is in principle no hope of settling that disagreement rationally. Here "rationally" is meant as opposed to by appeals to emotion, affection, decision, or commitment entered on some basis that has no objective standing between us.

But this impression is pertinent only if we take for granted that factual disagreement is different in that respect. And it seems rather that the cases are perfectly parallel.[11]

If I point out that your factual beliefs are in contradiction with mine, you could also retort "so much the worse for yours!" If I express an attitude or value judgment that pronounces negatively on what you are doing, you can retort "so much for your perverted sense of values!" These retorts are in perfect parallel, so far. In both cases, I have recourse. In both cases I will bring evidence to support my way of seeing the matter. In both cases, no evidence will have any weight for you unless it links significantly with something *you* think. There is a difference! In the case of factual disagreement, my evidence must link significantly with what you believe, while in the case of value disagreement, my reasons must link significantly with what you value. But the logical structure is parallel.

So there is a parallel between value disagreement and factual disagreement. Reason to despair in one case would be reason to despair in the other, but in fact there are (only!) precisely similar possibilities for dialogue in both.

> [Contrast 2] In the case of facts, we have a naturally shared access through perception, while there is no similar naturally shared access to whatever could settle value disagreement.

This point we can counter first of all by insisting that shared perceptions can be had only by people who have shared concepts, opinions, and beliefs. Any two people could be standing in front of the same thing, with light reflected from it impinging on their retinas. But this by itself will not give them the shared perceptual judgments that could settle a difference in factual opinion. The judgments they make spontaneously in response to this impinging light are heavily conditioned by the subjects' formation and previously conceptualized experience, their learning and the prior opinion that they bring to the occasion. Hence, once again, we see a parallel rather than a contrast.

But in addition, we can also attack this second putative contrast in another way. There is certainly a difference between seeing that e.g., the cat is on the mat, and seeing that the cat is graceful. In the latter case, the perceptual judgment is not straightforwardly factual, it is at least in part evaluative. However, a look at this second form, the perception of beauty, can lead us to a general insight, relevant here, that applies to both, and undermines that putative second contrast.

> In forming an estimate of objects merely from concepts, all representation of beauty goes by the board. There can, therefore, be no rule according to which any one is to be compelled to recognize anything as beautiful. Whether a dress, a house, or a flower is beautiful is a matter upon which one declines to allow one's judgment to be swayed by any reasons or

---

[11] See further my *Replies* in Monton 2007.

principles. We want to get a look at the object with our own eyes, just as if our delight depended on sensation. And yet, if upon so doing, we call the object beautiful, we believe ourselves to be speaking with a universal voice, and lay claim to the concurrence of everyone ...[12]

I'm not citing Kant to signal any agreement with Kant's views overall, and even less to call upon his authority. The point here that I want to emphasize comes in the second sentence: *there can be no rule to determine a value*. To this point I agree, and agree to it when extended to values other than aesthetic; but the emphasis must be on the word "rule."[13]

The remarkable tension Kant notes in aesthetic judgment is this: nothing would suffice ultimately to convince us of something's beauty except our personal sensual encounter with it, and yet, when we judge something to be beautiful, we "lay claim to the concurrence of everyone."

The very same thing is true of "factual" perceptual judgment. If I assert that something is red, when I see it, I also "lay claim to the concurrence of everyone." That much I think we will agree. But when it comes to this sort of judgment, e.g., that something is square, there too nothing can ultimately convince us of its truth except our personal sensual encounter with it. The encounter may need to be structured, with lighting and background systematically varied for example, and we may need convincing that the conditions are suitable for direct observation. But that is how it is with the question of beauty as well! And while much circumstantial evidence can be produced by measuring e.g., the scattering of reflected light, if that were ultimately in conflict with our senses then one would (to adapt Kant's words once more) decline "to allow one's judgment to be swayed by any reasons or principles." The physical or physiological theory of shape perception would have to bow to the evidence of the senses, and to be amended so as to restore empirical adequacy.

9.5   From a Liberal Epistemology to a Dilemma?

The false contrasts between factual and value judgments bring with them a putative dilemma for the views we can take of the status of value judgments

(i) "non-cognitivist:" value judgments are expressions of personal preferences or sentiments
(ii) "cognitivist:" value judgments are factual, and therefore susceptible to empirical or metaphysical investigation for their ratification or refutation.

---

[12] Immanuel Kant, *Critique of judgment* (trans. J. C. Meredith; eBooks@Adelaide 2004), First Part. Critique of aesthetic judgment. Section I. Analytic of aesthetic judgement. Book I. Analytic of the beautiful. Section 8. Cf. discussion by Alexander Nehamas, *Op. cit.*, beginning p. 47.

[13] Compare Jonathan Dancy's particularism in ethics, that continues the rather less-known strands of situationalist and existentialist ethics.

There may be sophisticated theories of value that go under those labels, but so blithely stated neither passes even first muster.

As to (i), it is easily seen that the logic of value judgment does not allow for equation with personal preference. We can quite consistently express personal preferences at odds with what we admit to be real values, or make negative value judgments about our own preferences, even while expressing a preference to maintain those very preferences. That is not logically inconsistent, though it may reveal e.g., weakness of will. But on view (i) this would be logically inconsistent.

And in (ii), the entire weight borne by this claim rests on what counts as empirical or metaphysical investigation. Speaking with common-sense, we surely can investigate empirically whether child labor or capital punishment is cruel, by first investigating child labor and then evaluating its consequences. If the evaluation of something as cruel or not cruel counts as empirical, then there is no bite to (ii). If it does not count, then we must refer back to our earlier discussion, with its conclusion that value judgments cannot be equated with their factual correlates.

Think here again, once more, about the difference between stating and expressing. I can state that I have certain values, or that someone else does; but I can also express my value judgments. Only the value judgments that I can express *as my own* play a role in my acting, judging, planning, practice. The point about expression, and the role that only the values, opinions, intentions, and aims we can express as our own can play a significant role in our practice, may perhaps abet the confusion with preferences, which can equally be either expressed or attributed. But this similarity provides no good argument for equating values, opinions, intentions, aims, and preferences.

9.5.1 Could There Be an Invariant Core of "Real" Objectivity?

Our evaluations of opinion, action, and methodology as instrumentally rational may be important for us, for the simple reason that they express our point of view, the stance we take, the goals we have adopted, the standards we set. Theoretically speaking, this may look idiosyncratic. Is any such value judgment invariant under all shifts in point of view?

Invariance is crucial to significance — in factual description. The suggestion here, in the above rhetorical question, is that invariance is crucial also for significance in value judgment. Recall Ayer's phrase "in so far as statements of value are significant …" The current suggestion could follow Ayer in form:

> in so far as statements of value are significant, they are the same in all points of view, and in so far as they are not invariant, they are not in the literal sense significant.

The "point of view" metaphor, closely related to visual perspective, which is illuminating in other ways, tends here to reinforce this suggestion, by the force of its analogy. Striking in the present context, about this particular metaphor, is its air of passivity. When I introduced the term "point of view," above, I included goals as well as opinions in one's point of view. But just because it is a visual metaphor, it may subtly subvert itself by removing goals from our attention. Different points of view are just different ways of seeing things, aren't they? So if different judges speak from within their own points of view, in pronouncing on someone's or something's rationality, do they convey any more than *how this looks to them*? We are not very far here from conceiving of value judgments as at heart purely factual, though perspectival — and perspectival factual judgments are just the sort that should give way to whatever they all have in common. But is there any warrant at all for expecting anything non-trivial to be common to all points of view? And if not, aren't we after all finding ourselves back in the debilitating relativism where value disagreement is impervious to reason?

In a way, this argument is not so different from the one we just finished. Its rhetoric temps us into a imagining ourselves as gods, viewing those points of view from outside, and thus seeing the values, goals, commitments, ideals involved in points of view as simply factual components of what different judges are like. To escape the tempter's lure, we need to remember that we are among the actors in this scene, and that the discussion can take on a radically different flavor if carried on in the first person, even when it pertains to shifts in point of view.

9.5.1.1 Plurality in the Scientific Enterprise

As a final effort then, to counteract this story of (as I see it) false contrasts and false dichotomies, let us turn directly to the putatively profounder challenge of relativism in the context of philosophy of science.

What I have in mind as telling example is the discussion, that followed on the publication of Kuhn's *Structure of Scientific Revolutions*, about the diversity in goals and standards among scientific communities. One reaction was to see that discussion itself as an attempted indictment of scientific objectivity. A more sympathetic reaction was to see it as showing up the irrelevance or inapplicability of any *a priori* standards of rationality. Philosophy of science began to admit the following view:

> Two scientific communities, even if starting with the same scientific background and receiving the same data, may nevertheless reach different conclusions and accept different theories, without thereby showing any defect of reason in either.

Such differing communities would also differ in their judgments of instrumental rationality. Is there any sense or way in which one could rationally and without bias judge the other? Should a community, that perceives such a disagreeing epistemic peer, change or lower its own convictions or confidence?

Suppose we place ourselves in the position of one such community. Imagine that we judge a certain procedure P to be instrumentally not rational, while in community X the same procedure P is judged to be instrumentally rational. Could we at the same time judge X's *evaluation of P as rational* as being itself rational?

If we say NO, then we mean that those who are in disagreement with us are to be dismissed as irrational, not epistemic peers after all. At first blush, that would seem to be a position not just arrogant but incapable of having any basis beyond arrogance.

But on the other hand, if we say YES, the conclusion must surely, again at first blush, be that judgments of [instrumental] rationality are so feeble, and so self-undermining once we look outside of our own bailiwick, that they are of no importance at all! In this imagined scenario, we would be saying "yes, X is instrumentally rational in taking procedure P as one that is instrumentally rational to follow, but that doesn't matter, because it isn't!"

The corollary applying to our own case would immediately follow: that we are instrumentally rational in our judgments of what is instrumentally rational provides no support whatever to our position. And so, mustn't we conclude, once again, that such judgments have after all no importance at all?

But how could any community, let alone one engaged in scientific inquiry, do without rational assessment of the value of means to ends?

With both NO and YES leading to such unacceptable conclusions, we face a dilemma, or so it seems.

9.5.1.2 The Dilemma's Incoherence

There is more insinuation than implication in this scenario. Consider the precisely same scenario, but with a factual disagreement between our community and community X about the empirical reliability or effectiveness of procedure P with respect to the same goal. That would be a factual rather than evaluative disagreement. If we regard X as rational, as an epistemic peer, would we then conclude that the issue at stake, the disagreement between us, is unimportant, or of no relevance to our own assessment?[14] Of course not.

But what about the more worrying scenario extended from the above, in which we and X disagree in our value judgment and agree on the facts?

Then our disagreement must derive from differing standards or goals, from what we think important and they regard as useless or laughable. But in that case, *by hypothesis*, we would not consider the issue at stake as unimportant either! The argument, as posed in the rhetorical questions above, placed us "between" two points of view, *suggested* that we imaginatively remain neutral between the two, and then asked us to conclude that the value judgments were of no importance. But "of importance" is a value-term, which makes tacit reference to a point of view when used — namely to the user's. It is logically incoherent to ask us *not* to consider whether an issue is important to us, or to community X, or to a specific other

---

[14] There are indeed arguments in the literature to the effect that if we meet disagreement with someone we count as an epistemic peer, we should not — on the basis of that alone — lower our own confidence in our views. There are also arguments to the contrary (Elga 2007, 2010; Kelly 2010). But we can leave that general issue aside: even if our confidence is not (or does not need to be) shaken by this encounter, it surely remains that it is to be taken seriously and requires a careful re-examination, and further exploration of the data to be had.

community, *while yet* demanding an answer to the question whether the issue is important. For our answer, if genuine, can only express, therefore reveal, what is important to us.

The idea that there are in principle many different scientific communities, all equally rational, all of them right in their judgments of scientific rationality, and yet disagreeing in those judgments, is subtly incoherent. For the phrase "equally rational" and "right" are both evaluative. Therefore the assertion in question, if genuinely used, would express the user's attitude, and his judgment would be incoherent.

It is a recurrently appealing and befuddling idea. There is no view from nowhere. In the above sort of argument couched in rhetorical questions we are asked to contemplate millions of points of view conflicting with our own, each of which could have been *our* point of view, namely if we had had any one of many different histories that shared a common beginning. In these different points of view we see different goals and very different opinions about what the world is like. Looked at "from above," which should we endorse? Would it not be totally *arbitrary* to endorse our own, the one we actually have, and say *we live here*, those goals are the ones which are *worthwhile*, that is what the world is like?

But by hypothesis that is the one we endorse! Endorsement reflects our own point of view, and is not endorsement if it doesn't. To say that we are arbitrary unless our own endorsement is point of view-free is to hold us to a logically impossible standard, asking us to judge without judging. If rationality and objectivity were identified through such a standard of non-arbitrariness, then they could indeed not exist; but we should not hanker after the logically impossible.

9.6   Conclusion

There is no escape from the normative, from value judgments, or from choices unconstrained by purely factual matters. The difficulties faced in developing epistemology, once this is faced, are serious, and all the more serious because they reappear at every level of reflection. But they are, if I am right, difficulties engendered by false dichotomies and false parallels, which drive seductive rhetorical questions.

References

Ayer, A. J. (1946). *Language, truth and logic* (2nd ed.). New York: Dover. (First published 1935)
Bashman, H. J. (2007, April 23). The trial judge as adversary on appeal. *Law.com*. http://www.law.com/jsp/article.jsp?id=1177059874555. Accessed 14 Jan 2013.
Chakravartty, A. (2012). A puzzle about voluntarism about rational epistemic stances. *Synthese, 185*, 37–48.
Elga, A. (2007). Reflection and disagreement. *Noûs, 41*, 478–502.

Elga, A. (2010). How to disagree about how to disagree. In R. Feldman & T. Warfield (Eds.), *Disagreement* (pp. 175–186). Oxford: Oxford University Press.

Foley, R. (1993). *Working without a net: A study of egocentric epistemology*. New York: Oxford University Press.

Giere, R. (1985). Philosophy of science naturalized. *Philosophy of Science, 52*, 331–356.

Giere, R. (1988). *Explaining science: A cognitive approach*. Chicago: The University of Chicago Press.

Kant, I. (1952[2004]). *Critique of judgment* (J. C. Meredith, Trans.). EBooks@Adelaide. http://ebooks.adelaide.edu.au/k/kant/immanuel/k16j/index.html. Accessed on 14 Jan 2013.

Kelly, T. (2010). Peer disagreement and higher order evidence. In R. Feldman & T. Warfield (Eds.), *Disagreement* (pp. 111–174). Oxford: Oxford University Press.

Laudan, L. (1990). *Science and relativism*. Chicago: The University of Chicago Press.

Monton, B. (2007). *Images of empiricism: Essays on science and stances, with a reply from Bas van Fraassen*. Oxford: Oxford University Press.

Nehamas, A. (2007). *Only a promise of happiness. The place of beauty in a world of art*. Princeton: Princeton University Press.

Putnam, H. (1982). Why reason can't be naturalized. *Synthese, 52*, 3–23; reprinted in Putnam, H. (1983). *Realism and reason, philosophical papers* (Vol. 3, pp. 229–247). Cambridge: Cambridge University Press.

Putnam, H. (1983). *Realism and reason, philosophical papers* (Vol. 3). Cambridge: Cambridge University Press.

Quine, W. V. O. (1977). Epistemology naturalized. In W. V. O. Quine (Ed.), *Ontological relativity and other essays* (pp. 69–90). New York: Columbia University Press.

Sheffy, Y. (1990). Unconcern at dawn, surprise at sunset: Egyptian intelligence appreciation before the Sinai campaign, 1956. *Intelligence and National Security, 5*(3), 7–56.

Teller, P. (2004). What is a stance? *Philosophical Studies, 121*, 159–170.

van Fraassen, B. C. (2002). *The empirical stance*. New Haven: Yale University Press.

Index of Names

Subject Index

W.J. Gonzalez (ed.), *Bas van Fraassen's Approach to Representation and Models in Science*, Synthese Library 368, DOI 10.1007/978-94-007-7838-2,
© Springer Science+Business Media Dordrecht 2014

Printed by Publishers' Graphics LLC